International
Launch Site Guide

International Launch Site Guide

Edited by Roy M. Chiulli

The Aerospace Press
El Segundo, California

The Aerospace Press
2350 East El Segundo Boulevard
El Segundo, California 90245-4691

Library of Congress Catalog Card Number: 94-079584
ISBN: 1-884989-01-2

Printed on recycled paper

Contents

Illustrations

Tables

Preface

The *International Launch Site Guide* is intended to provide payload planners general information useful in selecting candidate launch sites for a commercial payload. The data in this document have been collected from the most current public sources and through communication with the worldwide launch sites. The payload planner should contact the individual launch sites for specific data prior to final site selection.

This guide provides information on worldwide launch sites capable of launching commercial payloads, including those in the United States. Due to the large number of launch sites, this document presents general information only on selected sites. The launch sites in this document are those which have been historically active or are expected to be active in the near future.

The payload planner can best use the *International Launch Site Guide* in the following manner:

1. Consult Appendix 1, Commercial Launch Vehicles. Use the tables keyed to payload capacity and mission options to determine the commercial launch vehicles capable of performing the mission.

2. Once the set of candidate launch vehicles has been selected, consult Appendix 2, Worldwide Launch Sites and Launch Service Offerings. Use the tables keyed to launch vehicles and mission options to determine the launch sites able to perform the payload mission.

3. Consult the body of the document for those sites in Appendix 2 which are starred (*). A description of these launch sites and their facilities is provided.

4. Consult provided points of contact to obtain the specific information required for final selection of the launch site capable of performing the required payload mission. Additional information on launch sites can be obtained from the mailing addresses provided in Appendix 3.

The Aerospace Corporation intends that the *International Launch Site Guide* be used for reference only. Consult the launch site to obtain the best current and future information available. The Aerospace Corporation regrets any misinformation or outdated information included in this guide.

The *International Launch Site Guide* will be updated and published on a regular basis. The Aerospace Corporation solicits corrections and additional launch site information for inclusion in later publications.

The *International Launch Site Guide* is designed as a reference book. The information contained herein was developed for general reference only and therefore is not intended for specific technical application. The publisher and editors have used their best efforts to ensure the accuracy of the contents but assume no liability for errors or omissions.

An Overview of
The Aerospace Corporation

The Aerospace Corporation is a California nonprofit mutual benefit corporation. It operates a federally funded research and development center that is sponsored by the United States Air Force. Aerospace manufactures no products and does not compete with commercial companies, but serves the public good by providing technical advice which is unbiased by commercial concerns. Its overall mission is to enable its customers to achieve the full potential of space.

Since its founding in 1960, Aerospace has provided objective technical advice to the U.S. Government during the development of its national space program. It has also provided assistance to international agencies, including recently completed work on specifications for the Republic of China's ROCSAT program, assistance to the Centre Nationale d'Etudes Spatiale (CNES) with an analysis of the solid rocket motor for the Ariane 5 launch vehicle, and an assessment of digital signal processing techniques which might be used on next generation mobile personal communication satellites for the European Space Agency (ESA) and the International Maritime Satellite Organization (INMARSAT).

The corporation's areas of expertise include launch services (the subject of this handbook), launch vehicles, spacecraft, sensors, communications, ground systems, atmospheric physics, and other specialties. Aerospace has had engineering responsibility for more than 500 space launches, and currently oversees the operation of more than 90 operational satellites covering the full spectrum of space remote sensing, navigation, communications, and other missions.

The 2,200 members of the technical staff include world-class engineers and scientists (65 percent with advanced degrees; 25 percent with doctorates) in each space-related discipline. The staff provides a full range of in-depth coverage on all technical issues related to launch vehicle and space programs.

The Aerospace Corporation was part of the government/industry team honored in 1992 with the prestigious Collier Trophy for the "greatest achievement in aeronautics in America." The National Aeronautic Association awarded the trophy for the development of the Global Positioning System (GPS), a revolutionary satellite network providing precise navigational information anywhere on Earth. GPS has almost unlimited military and civil applications.

Worldwide Launch Sites

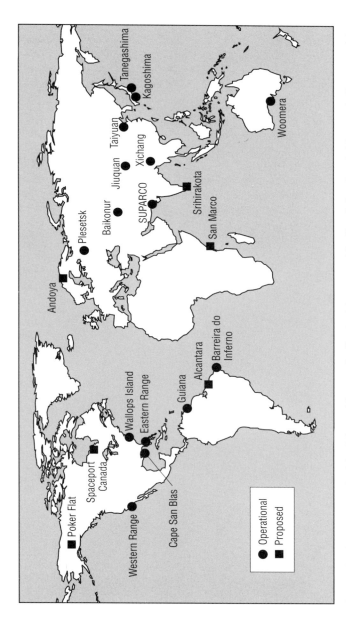

Selected Launch Sites

Launch sites presented in this guide are indicated here and in the following table, and discussed in detail in the main text. Additional launch sites are outlined in Appendix 2.

Table 1. Worldwide Launch Site Summary

Launch Site	Geographic Location	Launch Vehicles	Azimuths/ Availability	Mission Offerings
Australia				
Woomera Rocket Range	30.95° S 136.5° E	Optional	350 to 15° Continuous	Sounding rockets, low Earth orbit, Polar orbit
Brazil				
Alcantara Launch Center	2.17° S 44.23° W	SONDA III, IV, Veiculo Lancador de Satellites (VLS)	10 to 100° Not available	Sounding rockets, low Earth orbit (proposed)
Barreira do Inferno Launch Center	5.55° S 35.1° W	SONDA II, III, SONDA IV	14 to 145° Not available	Sounding rockets
Canada				
Churchill Research Range	57.7° N 93.8° E	Black Brant, Orion Sounding Rockets	Not available Not available	Sounding rockets, polar, low Earth orbits (proposed)
China				
Jiuquan Satellite Launch Center	40.6° N 99.9° E	Long March (1D, 2C, 2D)	135 to 153° Not available	Low Earth orbit
Taiyuan Satellite Launch Center	37.8° N 112.5° E	Long March 4	90 to 190° Not available	Low Earth orbit, sun synchronous
Xichang Satellite Launch Center	28.25° N 102.02° E	Long March 3, Long March 2E, Long March 3A	94 to 105° Not available	Geosynchronous transfer orbit
French Guiana				
Guiana Space Center	5.2° N 52.8° W	Ariane 4 (six versions), Ariane 5	-10.5 to 93.5° Not available	Low Earth orbit, geosynchronous transfer orbit, sun synchronous/polar orbit
India				
Srihirakota Range	13.7° N 80.2° E	ASLV, PSLV, GSLV	Not available Not available	Low Earth orbit, polar orbit, geosynchronous
Italy				
San Marco Equatorial Range	2.9° N 40.2° E	Scout	Not available Not available	Low Earth orbit
Japan				
Kagoshima Space Center	31.2° N 131.1° E	M-3SII, M-5	Not available Jan/Feb and Aug/Sept	Low Earth orbit, science
Tanegashima Space Center	30.4° N 131.0° E	H-I, H-II	Not available Jan/Feb and Aug/Sept	Low Earth orbit, geosynchronous transfer orbit, sun synchronous/polar orbit

Table 1. Worldwide Launch Site Summary (continued)

Launch Site	Geographic Location	Launch Vehicles	Azimuths/ Availability	Mission Offerings
Khazakhstan				
Baikonur (Tyuratam) Satellite Launch Center	45.6° N 63.4° E	Proton, Zenit, Energiya	Not available Not available	Low Earth orbit, geosynchronousos sun synchronous
Norway				
Andoya Rocket Range	69.3° N 16.0° E	Dart, Black Brant (To date)	Not available Not available	Sounding rockets, polar low Earth orbit (proposed)
Pakistan				
SUPARCO Flight Test Range	25.0° N 66.5° E	Centaure, Dragon	220 to 310° October to March	Sounding rockets
Russia				
Plesetsk (Northern) Cosmodrome	62.8° N 40.6° W	Soyuz, Molniya, Tsyclon, Cosmos	Not available Not available	Low Earth orbit, (high inclination)
United States				
Cape San Blas Launch Operations	29.7° N 85.3° W	Microstar	Not available Not available	Suborbital
Eastern Range[1]	28.5° N 80.6° W	Atlas II, Delta II, Lockheed Launch Vehicle, Taurus, Pegasus	37 to 112° Not available	Low Earth orbit, geosynchronous, geosynchronous transfer orbit, suborbital
Poker Flat Research Range	65.1° N 147.5° W	Sounding rockets	Not available Not available	Auroral and upper-middle atmosphere
Wallops Flight Facility	37.9° N 75.5° W	Conestoga, Orbex, Pegasus, sounding rockets	Not available Not available	Low Earth orbit, suborbital
Western Range[2]	34.6° N 120.6° W	Atlas II, Aquila, Conestoga, Delta II, Lockheed Launch Vehicle, Orbex, Pegasus, Taurus	147 to 281° Not available	Sun synchronous, polar orbit, suborbital

[1] Includes planned Spaceport Florida Launch Facilities at Cape Canaveral Air Station (CCAS)

[2] Includes planned California Commercial Space Launch Facilities at Vandenberg Air Force Base (VAFB)

Australia, Woomera Rocket Range

T he name Woomera is derived from an Aboriginal Native word given to a throwing stick used at arm level for extra spear throwing power. This early guided weapon launcher name applies to an area which was the location of Australia's and the United Kingdom's post World War II guided weapons development trials. Woomera specifically applies to the modern, well-serviced residential township located about 500 km by road northwest of Adelaide in the state of South Australia. The township is in the southeastern corner of a 127,000 km^2 area known as the Woomera Prohibited Area (WPA), about the same size as England. In the southeastern corner of the WPA (about 40 km from Woomera) are the Rangehead and Lake Hart launch vehicle sites within the Primary Trials Area known as the Woomera Instrumented Range (WIR). From a space applications point of view, the WPA and the WIR are managed by the Federal Government via

The Woomera Rocket Range has been host to a wide range of defense and scientific tests since 1948.

the Defense Department and are available for commercial use. The land is occupied and used by a native people, pastoral leaseholders (sheep farmers), and miners. However, the population is sparse. The whole WPA population is estimated at less than 1,500 people, with 1,200 located in Woomera.

Due to the availability of a large area of accessible land, the Woomera Rocket Range has been, since 1948, the host to a wide range of defense developments and scientific space experimental work. This includes guided weapons testing, sounding rocket testing, space capsule launch, and space capsule retrieval. As a result, there are considerable established resources and infrastructure available for commercial and scientific space and terrestrial applications.

The climate of this area is generally stable, warm, and dry, permitting year-round operations. The WPA is an arid zone area with sandhills, scrub, and low trees. The land is essentially flat with low, undulating hills in some areas and large dry salt pans. Access can be achieved by four wheel drive vehicles, helicopters, and small aircraft to nominated dirt landing strips. Major arterial roads and railways pass through the area allowing fast surface access.

Launch Site Description

Two launch sites are available at the Rangehead and Lake Hart. The Rangehead has small launch vehicle facilities, including significant established instrumentation. Lake Hart has concrete launch pads capable of supporting large launch vehicles. However, there are no permanent instrumentation or support facilities.

The Woomera Instrumented Range is an area of approximately 2,000 km^2 and includes the small launch vehicle facilities. The Rangehead has the following specifications:

- High operational availability
- Fully surveyed area
- Established safety procedures
- Security protection
- Solid motor and explosive storage facilities
- Small launch vehicle launch pads with launchers, reticulated power and water, bunkers, communications and air conditioned launch control facilities
- Reticulated electric power and water within the WIR
- A precision timing system accurate to 1 microsecond at a GPS receiver
- Optical instrumentation
- Radar tracking

- Tracking data system
- Local WIR communications system
- Fiber optic communications connecting into international trunk facilities
- Meteorological facilities
- Large launch vehicles and payload test shop/assembly facilities
- Hardware recovery facilities

Payload Accommodations

Payloads and launch vehicles can be transported to Woomera township via air, rail, or road transport. Ship port facilities are available at Adelaide. Transport from Woomera to the launching area (WIR) is via road. Test shop integration and assembly facilities are available at Woomera township or the Rangehead.

A large airport with a 2,440-m runway is available and usually operational at Woomera township. It can handle large freight aircraft (including the Lockheed C5A) and can also support aircraft that specialize in the air launch of space launch vehicles.

Woomera township has modern facilities to accommodate long- and short-term visitors supporting activities at the WIR. These include hotel and private house/apartment accommodations. Air, road, and rail transport is readily available. Phone, fax, and data communications are located in the town and at the Rangehead. Medical, hospital, schooling, recreational, and entertainment facilities are available.

Point of Contact

Australia Space Office
 P.O. Box 269 Civic Square
 Canberra, ACT. 2608
 Australia
 Tel: 61-6-276-1490
 Fax: 61-6-276-1567

Brazil, Alcantara Launch Center

The Alcantara Launch Center is a 620 km^2 site located near the town of São Luis (Maranhão State) on the Atlantic coast. Its formal opening took place in February 1990. This launch site benefits payloads with a gross 25 percent weight advantage over the launches from Cape Canaveral, due to its proximity to the equator. The site is in its final preparation for the launching of Brazil's first orbital launch vehicle, the Veiculo Lancador de Satellites (VLS) (Satellite Launching Vehicle), which is a four-staged, solid propellant rocket able to launch a 200-kg satellite into a 500-km circular orbit.

The maiden launch of the VLS can occur in the mid-1990s, carrying as payload a Brazilian Earth observation satellite, the Satellite de Sensoriamento Remoto/SSR-1 (Remote Sensing Satellite) followed by the SSR-2 a year later. The VLS is an international commercial launch vehicle: a joint venture between Avibras Aeroespacial SA of São José dos Campos, Brazil, and the China Great Wall Industry Corporation in Beijing, China, was formed in 1989. Named International Satellite Communication (INSCOM) Ltd., it markets Earth stations, communications satellites, and launch services. The consortium has already suggested the possibility of the Alcantara Launch Center as an equatorial launch site for the Long March vehicles. Commercial operations have also been discussed, since the center is specially suited for low Earth orbit launchings.

The launches of the Brazilian sounding rockets SONDA-I and SONDA-II have already been performed. In the second semester of 1994 the NASA mission DIP EQUATOR will be executed, in cooperation with Brazil's Instituto Nacional de Actividades Espaciais (National Institute for Space Research). This mission will perform scientific experiments on the equatorial ionosphere through 33 sounding rockets, including Black Brant, Nike-Tomahawk, and Nike-Orion.

Alcantara has two distinct seasons. Rain falls from January to June, with a mean precipitation of 200 mm a month, peaking up to 350 mm around April/May with a constant, mild rain. The drought months follow with a mean precipitation of 20 mm a month, mostly from very sparse summer rain. The temperature fluctuates about 26.5 °C all year round, just as the relative humidity clings to the 85 percent mark. The predominant wind is northeastern in both seasons, blowing with a mean velocity of 2.5 m/s, reaching peaks around 4.0 m/s during the third quarter.

Launch Site Description

Azimuth
Launching azimuths ranging from 10 to 100°.

Preparation and Launch Pad
These buildings include the preparation facility for vehicles and payloads, launch pad, and advanced control (blockhouse).

Operating Systems
The operating systems available in Alcantara are: radar system, telemetry system, control center, and meteorology facilities.

Radar System
The radar system consists of Adour and Atlas-Thomsom Radar Types. The Adour has a 3-m-diameter antenna, frequency range from 5,450 to 5,825 MHz, peak power of 250 kW, and average power of 250 W. The Atlas has a 4-m-diameter antenna, frequency range from 5,450 to 5,825 MHz, peak power of 1 MW and average power of 1 kW.

Telemetry System
The telemetry system operates in S and P bands. The S band has 10-m-diameter antenna, frequency range from 2,200 to 2,290 GHz, four receivers (two double chain carriers), decommutation PCM, PAM, and FM. The P band has a small size antenna, frequency range from 215 to 260 MHz, and two receivers (one double chain carrier).

Control Center
The control center includes all facilities to control the operations during the countdown.

Meteorology Facilities
The meteorology facilities are designed for altitude and surface meteorological observations, as well as follow-up of frontal systems and mesoscale required for weather forecast.

Satellite Control Station
This facility consists of a ground station, a satellite preparation complex, a satellite filling and assembly building, a chemical laboratory, and storage for pyrotechnics and propellants.

Future Implementations
In addition to the actually occupied area of 180 km^2, there is a provision for expanding into an area of 340 km^2 to implement future installations such as the ones for liquid propellant vehicles.

Payload Accommodations

The payloads can reach Alcantara by air, land, or sea transportation. The center has a 2,500-m by 45-m runway. It allows operations of large size aircraft, in the class of B-707, B-747, and equivalents.

There are special buildings for storage and preparation of vehicles up to 50 tons, as well as facilities in final stage of construction for payloads weighing about 2,000 kg.

The standard electrical power is available in 220/380V, 60 Hz. There is an emergency power system available for supplying power as demanded for specific sectors.

A data processing network, consisting of a central SOLAR 16/85 computer located at the data processing center and two SOLAR 16/65 computers located at radar stations, assures accuracy and expedience to operations.

A time generation and distribution system assures general time synchronization for the center.

There are accommodations for 188 persons during a launching campaign.

A management support system is available including interphone networks, specialized telephones, and operating signaling. The center provides for both national and international telecommunications, such as long distance calls, fax, telex, data link, and the like.

Points of Contact

General Coordination

Departamento de Pesquisas y Desenvolvimento
   Esplanada dos Ministérios–Bloco M
   Edifício Anexo do MAer, 3°Andar
   70.045-900–Brasília–DF Brasil
   Fax: 55-61-2246123

Technical and Operational Information

Centro de Lançamento de Alcantara
   CEP: 65.250-000–Alcantara–MA Brasil
   Fax: 55-98-2111069

Brazil, Barreira do Inferno Launch Center

The Centro de Lançamento da Barreira do Inferno (CLBI) (Barreira do Inferno Rocket Range) is part of the Brazilian Space System, devoted primarily to suborbital space operations. Sponsored by the Ministry of Aeronautics, the CLBI works in close cooperation with the Instituto Nacional de Atividades Espaciais (National Institute for Space Research) and the local university (Universidade Federal do Rio Grande do Norte). It provides management, testing, launching, tracking, and data acquisition operations, to both Brazilian and foreign sounding rocket programs.

On December 15, 1965, with the launching of a Nike-Apache sounding rocket, operations at the CLBI were initiated. Since then several international projects have been carried out, involving, for instance, the National Aeronautics and Space Administration (NASA), the Air Force Geophysics Laboratory (AFGL), and the Deutsche Forschungsanstall fur Luft-und-Raumfahnt (DLR).

Since 1979, as a result of an agreement between the European Space Agency (ESA) and the Brazilian Space Agency (Agência Espacial Brasileira), the range facilities were improved to enable the CLBI to support the Ariane Program. This support consists of radar and telemetry data acquisition, real-time data processing, and transmission to Centre Spatial Guyanais and Centre Spatial de Toulouse.

The CLBI is located in northeast Brazil, 17 km south of Natal City (05°55' S./035°09' W.). The range has an area of approximately 18 km^2, bordered on the east by 6 km of seashore. The CLBI was surveyed with utmost precision, resulting in well-defined benchmarks for all stations.

The range location affords excellent safety conditions and allows broad launching azimuth availability, from 14 to 145°.

The CLBI offers scientists equatorial latitudes, quick response, and low cost services.

The climate is warm and dry almost all year. The short rainy season brings moderate precipitation. Surface winds blow moderately and constantly from the southeast.

Launch Site Description

Launch Pads

Pad 1 is a circular concrete area, 24-m-diameter, equipped with a Universal Rail Launcher. Pad 2 is a rectangular concrete area (20 by 6 m), equipped with a MRL 7, 5K launcher. Pad 3 is a circular concrete area, 20-m-diameter, equipped with a Nike-type launcher, which is also used to fit the Super-LOKI launcher adapter. Pad 4 is a circular concrete area, 10-m-diameter, equipped with a LOKI tubular launcher and test rocket launcher. Pad 5 is a rectangular concrete area (30 by 5 m), used to place mobile launchers. Pad 6 is a rectangular concrete area (50 by 10 m), equipped with the SONDA IV assembly tower and launch platform.

Blockhouse

This is a circular structure, partly underground, built in reinforced concrete, with 12 protected windows allowing direct visual observation of the pads.

Pyrotechnics Assembly Building

This is a 28 m^2 building constructed and equipped in accordance to Safety Regulations.

Storage Buildings

There are two storage buildings, one specific for rocket motors storage (93 m^2) and another for nonhazardous materials (500 m^2).

Instrumentation

Radar System I

The Bearn is a C-band, scanning tracking radar, with the following characteristics: 4-m-diameter antenna, frequency range 3,450–5,825 MHz, and 1 kW peak power.

Radar System II

The Adour II is a C-band, scanning tracking radar, with the following characteristics: 3-m-diameter antenna, frequency range 5,450–5,825 MHz, and 250 kW peak power.

Telemetry System

This is a S-band system with the following technical data: 10-m-diameter antenna, frequency range 2,200-2,300 MHz, 6 receivers, decommutation PCM, PAM, and FM-FM.

Data System

Tracking management and control are accomplished by two SOLAR 16/85 computer systems interfacing the radars and telemetry stations. In addition to controlling trajectory data acquisition and tracking, the computer systems furnish recorded trajectory data.

The telemetry ground station has dedicated equipment that performs internal data processing.

Time Generation System

This system assures, for all operational stations, general time synchronization and distribution of coded time.

Communications System

This system provides intra-stations and external voice, signaling, and data communications. Intra-station communications are assured by a cable network, whereas external communications are assured by the Brazilian Telecommunications Company (Empresa Brasileira de Telecomunicações).

Command/Destruct System

The CLBI operates a 200 W output power command/destruct system.

Management and Control

These functions performed at the Control Center are supported by a signaling system that displays the status of the stations and enable the Go/No-Go circuits.

Payload Accommodations

Payload Assembly Building

This is a four-room building with a total available area of 180 m^2. Grounding system, communications, temperature control, and many other support systems and equipment are also available.

Vehicles Assembly Building

This is a 120-m^2 building constructed in reinforced concrete. Ground support equipment, such as hoists, dollies, forklifts, air compressors, etc., are available for rocket checkout and assembly.

Power Supply System

Redundant power supply systems are available for all the stations from commercial and locally-generated power. The CLBI power plant consists of two 330 kVA generators (principal and backup).

User Support Services

In addition to operational services, the CLBI offers two classes of user support services:

- **Standard Services:** pre-operational support (authorizations, official documentation, reservations, etc.), customs clearance, local transportation of material and personnel, material storage, workshop support, and health assistance.

- **Optional Services:** lodging, meals, car rental, material acquisition, equipment repairing, aircraft support (parking, refueling, and maintenance), and air cargo.
Payload recovery service is also available.

Airport

Natal Airport (Augusto Severo International Airport) is suitable for operations for all types of transport aircraft. It provides parking, refueling, and flight operations services. Customs and health authorities are also available.

Points of Contact

General Coordination

Departamento de Pesquisas y Desenvolvimento
    Esplanada dos Ministérios–Bloco M
    Edifício Anexo do MAer, 3°Andar
    70.045-900–Brasília–DF Brasil
    Fax: 55-61-2246123

Technical and Operational Information

Centro de Lançamento de Barreira do Inferno
    RN 063, KM 11
    Caixa Postal 640
    59150-000–Natal–RN Brasil
    Fax: 55-84-2114226

Canada, Churchill Research Range

Proposed Spaceport Canada

T he Churchill Research Range is a 700,000 km^2 site (including the majority of Hudson Bay) which was built during 1956 and 1957. The site is approximately 900 km north of Winnipeg, Manitoba, and was constructed to support Aerobee and Nike Cajun-class sounding rocket firings. The northern location provides a high auroral concentration (up to 265 days per year) and is well suited for the study of arctic ozone depletion.

The site has been under the control of the U.S. Army, the U.S. Air Force, and Canada's own National Research Council (NRC) since the range first opened. In 1980, the NRC delegated responsibility for the Churchill Research Range to the Canada Centre for Space Science's Facilities Branch. The range was closed in 1984, whereupon much of the support equipment was removed from the site. In 1989, the Manitoba Aerospace Technology Program completed a study on the feasibility of reactivating the range and using the range for polar orbital launches.

In March and April 1989, two Black Brant 10 and two Nike-Orion sounding rockets were fired from an existing enclosed launcher and tracked using mobile equipment. These were the only launches since 1985.

Akjuit Aerospace, Inc., Winnipeg, is planning to refurbish and upgrade the facilities at the Churchill Research Range for a commercial launch service operations venture. The site will be redesignated as Spaceport Canada. Initially, the site will be offered for sounding rocket services. Later, site expansion is planned to provide a satellite launch facility for the launch of small payloads into polar orbit. A lease agreement allows Akjuit to occupy the Churchill Range in mid-1994. Sounding rocket services could follow in late 1994. Satellite launch services are proposed for late 1996.

Launch Site Description

The site consists of four launch pads: Pads 1, 3, 4A, and 7. Pad 1 consists of a Universal Launcher that can handle vehicles up to 18-m-long and weighing up to 4,500 kg. The enclosed launch bay is temperature controlled. Vehicles using this facility include the Nike-Tomahawk, the Black Brant, the Taurus-Orion, and the Nike-Orion. Pad 3 consists of an Aerobee tower designed for the Black Brant vehicle configurations. Pad 4A consists of two Arcas launchers. Pad 7 consists of an enclosed Auroral launcher which can handle vehicles up to approximately 12-m-long and weighing up to 1,800 kg.

Vehicles using this facility have included the Black Brant, Nike-Tomahawk, and Nike-Orion.

Points of Contact

Manitoba Aerospace Technology Program
    500-15 Carlton Street
    Winnipeg, Manitoba R3C 3H8
    Tel: 1-204-945-2030
    Fax: 204-957-1793/945-1354
Canadian Space Agency
    Place Air Canada
    500 René-Lévesque Boulevard West
    Montreal, Quebec H2Z 1Z7
    Canada

China, Jiuquan Satellite Launch Center

T he Jiuquan Satellite Launch Center (JSLC) is located in north-central China in the Gansu Province, near the Mongolian border on the southern edge of the Gobi desert (about 1,600 km west of Beijing near the Soviet border) north of the Jiuquan city in Kansu province. The remote site location (altitude <1,000 m) was chosen due to the nature of early test launch activity and low population density. Launch corridor azimuths are constrained from 135 to 153°. Typical missions are south-easterly into orbits between 57 to 70° inclination to avoid overflying the Confederation of Independent States (CIS) and Mongolia.

The JSLC is the oldest of China's launch facilities and comprises two launch pads. The primary support area is about 90 km south of the main base area Dong Feng (the East Wind) where a hotel is located. A Technical Center for launch operations support is located about midway between Dong Feng and the launch area. The center is served by Airport No. 14. There are no scheduled flights to the JSLC. The only alternative is to fly to Lanzhou (capital of Gansu Province) and then take the train (20 hours), but this requires a Chinese escort.

The JSLC was established as China's primary rocket test facility in the early 1960s. By 1964, launch facilities were in place to support the first launch of a ballistic rocket. In April 1970, China's first satellite was launched from the JSLC. All Chinese space launches originated from the JSLC through 1983. The primary domestic and commercial mission for this center is to provide low Earth orbit launch capability including all recoverable satellite missions. The first commercial recovery mission was launched in August 1987. In October 1992, a Swedish Space Corporation scientific satellite was launched as a commercial piggyback (free-flyer) satellite with China's domestic FSW-1 payload on a Long March 2C.

The JSLC is used primarily for recoverable Earth observation and microgravity missions. Because of its geographical constraints, increased commercial activities are focused at China's other launch sites.

Launch Site Description

The Technical Center where the launch vehicle is assembled is about 30 km from Dong Feng. The JSLC launch facility consists of two launch pads about 300 m apart with a shared service tower. A large rail-mobile gantry provides service to both launch positions. Each launch position comprises a launch stand, aperture, and single exhaust duct. Each position is attended by a fixed service tower with rotatable service platforms. There are no provisions for environmental protection of the launch vehicle at either pad; the mobile gantry provides this service. During the on-pad integration operations, crane service is provided by the mobile gantry for launch vehicle and payload lift operations. As of 1993, the Long March 2C and 2D were the only space vehicles performing launch operations from the JSLC.

Payload Accommodations

Payloads are checked out in the BM building, newly built for solid rocket motor checkout, with about class 300,000 air quality. A clean tent should be brought to the site if required. Payloads may be flown to Airport No. 14 and reloaded on a train for the trip to the Technical Center railway station. The payload then is reloaded for the trip to the BM building. The train ride from Airport No. 14 to the Technical Center takes 2 hours and is very smooth.

The power is 220/380 V AC. The three-phase system has only four wires and the ground resistance is high due to the desert climate.

There is a good clean room (better than class 300) at the top of the gantry tower that protects the satellite very well. This air quality is maintained even when there is a strong wind blowing across the launch pad.

An IDD telephone line, going through a Chinese geostationary satellite to reach Beijing and from there via Intelsat to the rest of the world, is provided. A similar line is provided in the blockhouse. Due to the double hop, the voice quality is not high. However, the fax connection works well. The hotel suites are satisfactory, but the hotel is old. The food is good. While there is little entertainment available for guests, the Chinese hosts arrange social events and outings.

Points of Contact

China Great Wall Industry Corporation (GWIC)
    21 HuangSi DaJie
    Xicheng Qu
    Beijing, 100011
    P.R. China
    Tel: 861-837-2708, 861-837-1682
    Fax: 861-837-3155, 861-837-2693
GW Aerospace, Inc.
    21515 Hawthorne Blvd. #1065
    Torrance, CA 90503
    U.S.A.

China, Taiyuan Satellite Launch Center

The Taiyun Satellite Launch Center (TSLC) is located in the Shanxi Province in northeastern China. The launch site location (altitude less than 1,500 m) was chosen initially to provide China with domestic sun synchronous and polar satellite launch capability. The launch corridor azimuths are constrained to be from 90 to 190°. The site is currently offered as a commercial launch facility for the Long March 4 launcher.

The historic details of the TSLC have not been released. The site became an operational satellite launch facility in September 1988 by supporting the inaugural launch of the Long March 4 for remote sensing, meteorological, and reconnaissance missions. The second launch occurred in September 1990.

Launch Site Description

The TSLC launch facility is patterned after the Xichang Satellite Launch Center Site 1. The launch pad comprises a launch mount, exhaust aperture, and a single below-grade exhaust duct. The launch position is serviced by a fixed service tower with rotatable service platforms. A large service crane is mounted atop the tower that is utilized for lift operations during the onpad integration process. Lightning protection is provided by lattice-type conductor towers around the perimeter of the site. As of 1993, the Long March 4 was the only active launcher associated with the TSLC.

The Taiyuan Technical Center for launch vehicle and payload processing has not been publicized to date. The launch vehicle processing is probably the same as conducted at the other of China's launch centers. Payload processing is expected to be at a level that will accommodate commercial launch operations. As of 1993, no commercial launch operations have been conducted at the TSLC.

Point of Contact

China Great Wall Industry Corporation (GWIC)
    21 HuangSi DaJie
    Xicheng Qu
    Beijing, 100011
    P.R. China
    Tel: 861-837-2708, 861-837-1682
    Fax: 861-837-3155, 861-837-2693

China, Xichang Satellite Launch Center

The Xichang Satellite Launch Center (XSLC) is one of the launch centers included in the China Satellite Launch and Tracking Control General. The main purpose of this center is to launch spacecraft such as broadcasting, communication, and meteorological satellites using Long March launch vehicles. The XSLC is located in a nearly sub-tropical mountainous region (altitude 1,826 m) of the Sichuan Province of southwestern China. The XSLC is the most southerly of China's launch centers and can be compared in geographic latitude to the U.S. Cape Canaveral. It was selected from a list of 16 sites in a search for a more favorable geosynchronous mission base than Jiuquan.

The center comprises two launch sites along with support facilities. This area is surrounded by agricultural activity and occupied dwellings that co-exist with the launch site activities. This site was chosen to provide China with geosynchronous launch access for domestic and commercial launch operations. The XSLC headquarters are located in the city of Xichang about 65 km south of the launch site. The Xichang Airport is about 50 km from the launch site. The runway of the airport, at 3,600-m-long, is capable of landing jumbo aircraft such as C-130s and Boeing 747s.

The Xichang location provides a 28° inclination launch corridor (azimuths between 94 to 105°) free of overflight restrictions. Construction was started in the canyon site at the foot of Mt. Liang Shan on the first launch pad, Site 1, in 1978 and was completed for the inaugural launch of the Long March 3 in January 1984. Following China's entry into the commercial launch services market, additional payload handling facilities were constructed to support foreign spacecraft handling operations. In addition, modifications were also performed to the onpad service structure to provide a clean room environment for commercial payload mating operations.

Following successful domestic geosynchronous satellite launches, the first commercial launch, Long March 3/Asiasat, occurred on April 7, 1990. As the size and mass of commercial launches increased, a second pad, Site 2, was designed for a follow-on Long March 2E vehicle. This second pad is located about 400 m from Site 1 and was constructed in about 14 months. Site 2 was available for the inaugural commercial demonstration Long March 2E launch in July 1990. More recently, in February 1994, Site 2 supported the inaugural launch of the Long March 3A.

The climate of the Xichang Region is subtropical with a temperature range from 7 to 25 °C (the yearly average temperature is 16 °C). Wind

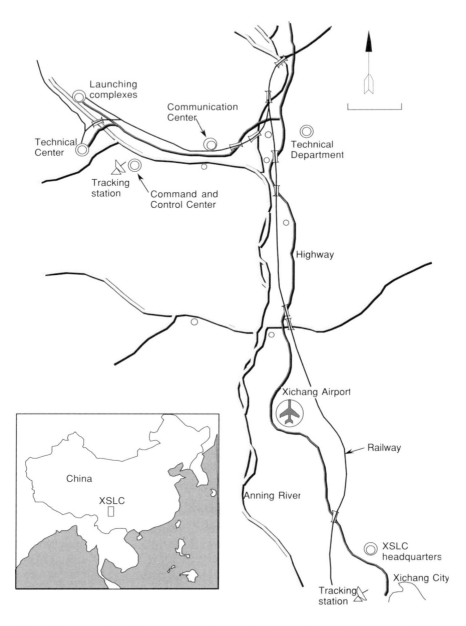

The Xichang Satellite Launch Center is located in southwestern China (Sichuan Province) about 50 km from Xichang Airport.

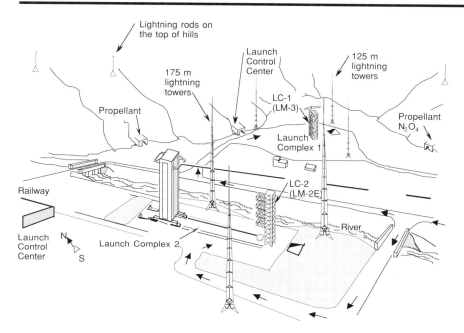

Launch Complex 1 (LC-1) and Launch Complex 2 (LC-2) are used for launch of the Long March 3 and Long March 2E vehicles, respectively.

speed in this region is very low at all seasons. The wet season is from June to September, while the dry season is from October to May.

Launch Site Description

Technical Center

The Technical Center consists of the launch vehicle preparation facilities and the payload preparation facilities. It is 2.2 km from the launch complex. The launch vehicle and the payload are processed, tested, checked, assembled, and stored in the Technical Center. The launch vehicle preparation facilities consist of the Transit Building (BL1) and the Testing Building (BL2). A railway branch leads directly into the Launch Vehicle Preparation Building, with a turnaround line of 260 m. A loading dock and freight house are available. The payload preparation facilities include the Nonhazardous Operation Building (BS2), the Hazardous Operation and Fueling Building (BS3), the SRM Checkout and Preparation Building (BM), and the SRM X-ray Building (BMX).

Launch Areas

Facilities in the launch area mainly consist of the existing LM-3 launch complex LC-1 and the LM-2E launch complex LC-2.

Mission Command and Control Center

The Mission Command and Control Center (MCCC) is 7 km southeast of the launch area. The building is divided into two parts: the command and control hall, and the computer room. The command and control hall consists of the command area and the safety control area. Around the hall are operation rooms and offices. There is a viewing room on the second floor where visitors can see the displays on the large TV screen. Closed-circuit TV sets are also equipped in the visitors' viewing room.

Launch Operations

The launch operations procedures described in this section are typical. Users may set up their own operation plan and procedure. However, the final operation plan and procedures should be compatible with that of the launch vehicle and agreed upon by the parties. Normally, the preparation and operations on the launch vehicle require 30 to 45 working days, while the spacecraft requires 45 to 60 days. After the launch vehicle and the spacecraft have been handled in accordance with the independent preparation procedures of each, joint operations are performed.

The XLSC provides services for transportation, propellant storage, gas supply, RF relay link, and communications. For foreign spacecraft launches, satellite communications including telephone, telex, fax, and data transmission are used to link the XSLC to the user's country. Dedicated duplex communication links can be established via two paths between the XSLC and the user's country. During the mission, only one path is used while the other path serves as a back-up link. Communication service includes:

- Audio Communications: Telephone and facsimile communications can be established by using the secretary's console and facsimile machine via the dedicated circuitry of the XSLC.

- Data Communications: Transmission rate of international data communications will be 2400 bps. The maximum rate is 4800 bps.

- Telex: Communications can be established via the Telegraph Automatic Switching Network for international communication.

Point of Contact

China Great Wall Industry Corporation (GWIC)
21 HuangSi DaJie
Xicheng Qu
Beijing, 100011
P.R. China
Tel: 861-837-2708, 861-837-1682
Fax: 861-837-3155, 861-837-2693

French Guiana, Guiana Space Center

The Guiana Space Center or Centre Spatial Guyanais (CSG) is located on the Atlantic coast of French Guiana, South America. This coastal region extends more than 30 km north of the town of Kourou (altitude <200 m) with launch facilities situated about 20 km from Kourou. The CSG is located about 65 km north of Guiana's largest city, Cayenne, which is the primary logistical support base providing air and port facilities. Although the launch facilities are within a few kilometers of the ocean, the salinity of the coastal waters has been diluted by fresh water from the Amazon River, which reduces the corrosive effects on launch facility hardware and launch vehicles. The launch facilities are located in an area that is free of earthquakes.

The near-equatorial location of the CSG is ideally suited for a launch facility because of the greater velocity gain due to the Earth's rotation compared with other commercial launch sites located at higher latitudes. Guiana's coastline allows launches into equatorial and polar/sun synchronous orbits. The launch azimuths are restricted to between −10.5 and 93.5°. The low population of Guiana and a launch range that is essentially a broad ocean area provide nearly unrestricted access to space for most mission orbits.

Though the Guiana Space Center is owned by CNES (the French national space agency), the launch site is made available to European Space Agency (ESA)/Arianespace under a French government agreement which guarantees access to the ESA-owned launch facilities. CNES chose the Kourou, French Guiana, site for its national space launch center following closure of its launch center in Hammaguir, Algeria, in 1964. By 1966, the decision was made to transfer European space launch operations to the French-owned facility under agreement with the CNES. In 1970, the first satellite was launched from Kourou on a French Diamant B launch vehicle. By 1973, Europe's Europa II program was terminated, followed by the termination of the Diamant program in 1975. In 1976, the ESA decided to utilize the Kourou facility for the Ariane launch program. Modifications were performed on the deactivated Europa II launch complex and the site was redesignated as the Ensemble de Lancement Ariane 1 (ELA 1). Launch processing at ELA 1 required the onpad integration of the launch vehicle and payload. ELA 1 became operational in December 1979 with the first launch of an Ariane 1. The launch capacity was limited to five or six per year. Follow-on Ariane 2 and 3 vehicles were also launched from ELA 1 from 1984 through 1989. In June 1991, ELA 1 was rendered unserviceable (deactivated) with the intentional razing of the service tower.

Construction of the Ensemble de Lancement Ariane 2 (ELA 2) began in mid-1981 and was completed in 1985; the Ariane 4 launcher was not compatible with the ELA 1 facility. The ELA 2 design objective was to provide a more efficient, flexible, and less time-consuming method of launch vehicle processing. The Ariane 4 undergoes offpad, partial integration before being transferred vertically to the launch position using a mobile launch table. The Ariane 3 launcher was used as the qualification and verification vehicle for ELA 2 which began in March 1986. The first launch of an Ariane 4 was conducted in June 1988.

The Ensemble de Lancement Ariane 3 (ELA 3) is a third generation facility that will support launch operations of the Ariane 5 into the next century. Site construction was started in 1988 and represents the largest project to be approved by ESA to date. The ELA 3 was designed to be a simplified and resilient pad facility. Launch processing operations for the Ariane 5 utilize the offpad, full integration technique. The vehicle is processed vertically in offsite facilities and then transported to the pad fully integrated. The first launch of the Ariane 5 from ELA 3 is expected in 1995.

Launch Site Description

Launch Site ELA 2 (Ensemble de Lancement Ariane 2)

The ELA 2 facility is divided into two interconnecting zones described as the Launch Zone and the Launcher Preparation Zone. This facility arrangement was used as a second generation improvement in overall Ariane operational launch capability which provides for the offpad, partial integration of the Ariane launcher. This processing method reduces the onpad prelaunch timeline and allows parallel launch vehicle processing. This was accomplished by locating the launch vehicle processing facility (Preparation Zone) in close, but safe, proximity to the launch facility.

The pad facility, the Launch Zone, is comprised of a massive subsurface concrete support structure with a surface level launch aperture connected to two angled exhaust ducts. A fixed, 74-m-tall umbilical tower, located 12 m from the pad aperture, provides service connections for the onpad operations. The third stage cryogen loading is supplied by the umbilical tower. The pad is also served by an 80-m-tall mobile service gantry that is connected to the pad by rails. During onpad operations, the upper portion of the service gantry mates with the upper portion of the umbilical tower forming a sealed, class 100,000 environmental shelter for launch vehicle and payload service operations. During launch, the service gantry is stowed in a parking position 80 m from the pad. The pad is also served by a dual set of tracks 950-m-long that interconnect the pad to the Launcher Preparation

Zone. The tracks are utilized by the launch vehicle mobile launch table (MLT) during prelaunch and postlaunch operations. An air-bearing turntable is located halfway between the pad and the Launcher Preparation Zone to provide a bypass capability for multiple MLT transit operations. A set of branch tracks extends from the turntable to an offline MLT parking position that is normally used to stow the second of two MLTs when not in use. Ground launch support structures provide the primary lightning protection for the launch facility.

The ELA 2 MLT serves as the launch support mount during launch. In addition, the MLT provides the transport capacity for the launch vehicle to the launch pad after the launch vehicle processing has been performed in the Launcher Preparation Zone. The MLT is a compartmented, steel structure 13- by 13- by 4-m high, weighing approximately 500 tons. Two MLTs were built to conduct parallel (simultaneous) launch processing operations. Initially, one of the two MLTs was modified with an additional 7 m elevated structure to support the Ariane 3 launch program. (The Ariane 3 was used for qualification and verification launches for ELA 2 prior to the start of the Ariane 4 launch program.) The additional height was needed to provide onpad umbilical tower compatibility with pre-existing higher-level Ariane 4 service connections. The modified MLT was designed to be reconfigured for the Ariane 4 following the end of the Ariane 2/3 launch program. During prelaunch transport operations, a rail tractor provides the MLT with locomotive power. The transit time to the pad is less than 1 hour.

The Launcher Preparation Zone, located 950 m from the Launch Zone, comprises three contiguous sections that provide for the receiving, storage, checkout, and prelaunch integration of the Ariane 4 launcher. The buildings are designated as the Destorage Hall (receiving and storage), Erection Hall (pre-integration launch vehicle stage preparation), and the Assembly Dock (Ariane integration).

Launch Site ELA 3 (Ensemble de Lancement Ariane 3)

The ELA 3 facility is divided into three functional areas: the Launch Zone, Preparation Zone, and Solid Propellant Booster Zone. This third-generation launch support and launch facility designed for offpad, full integration virtually eliminates onpad processing time, which in turn reduces the requirement for onpad service structures.

The Launch Zone contains a pad that was designed to be simple and resilient, without major service support structures; a "flat pad" or "clean pad" concept. The pad aperture is serviced by an underground three-way exhaust system. The pad is connected to the Preparation Zone by a 2.8-km-long dual set of rail tracks that is utilized by the MLT during launch vehicle transport operations. The only primary onpad structure is located immediately adjacent to the MLT onpad prelaunch position which provides protection

25

for propellant and other service systems that are connected to the MLT. A large water storage tower is the tallest major structure onsite that supplies the water-deluge noise-suppression system during launch. Unlike the ELA 2 launch site, there are four lightning conductor towers around the ELA 3 launch position. The use of towers is due to the absence of onpad service structures which normally would provide lightning protection for the launch vehicle. Provisions for onsite storage and transfer systems for the liquid oxygen and liquid hydrogen core stage propellants are connected to the onpad service support structure. The Ariane launch pad was also designed to serve as the qualification static test position for the core-stage Vulcain engine prior to the start of the Ariane launch program.

The ELA 3 MLT is a massive, 900-ton, rail-mobile structure that serves as the primary launch vehicle transport and launch support platform. As was the case at ELA 2, two MLTs provide parallel prelaunch processing capability. An umbilical tower has been added to the ELA 3 MLT that, by design, eliminates the need for a fixed onpad tower. In addition, the requirement for an onpad service gantry is eliminated due to the Ariane 5 offpad, full integration processing. The Ariane 5 vehicle design requires three launch pad compatible exhaust apertures within the MLT which are offset opposite the integral umbilical tower position. The solid motor apertures, located on each side of the core tank aperture, are uniquely configured to serve as receptacle positions for the pallet-mounted solid motors during the integration process. From the start of the vehicle-to-MLT integration process and until launch, the MLT umbilical tower is an active interface element in the Ariane checkout, monitoring, and launch countdown.

The Preparation Zone comprises those facilities in which the primary launch vehicle integration is performed. Initial integration will be performed in the Batiment Integration Lanceur, the Launcher Integration Building. The final prelaunch integration will be performed in the Batiment Assemblage Final, the Final Assembly Building. Each facility is interconnected by the MLT rail system.

The Solid Propellant Booster Zone is the area in which the solid boosters are processed and vertically assembled in a building containing two integration cells designated the Batiment Integration Propulseurs (Boosters Integration Building). Each booster is assembled on special mounting pallets that are transportable using air-cushioned rail transporters. This facility is safely located about 3 km from the Launcher Integration Building. The Boosters Integration Building is also connected by rail to the Guiana Propellant Plant and the Solid Booster Test Facility.

Construction of the Ariane 5 facilities began in 1988. ELA-3 will be used for the development and qualification of the Ariane 5's core and Vulcain main engine. Completion of acceptance testing of the Ariane 5 facilities is scheduled in early 1995, with the first launch in late 1995.

Points of Contact
BP-726, F-97387 Kourou Cedex
  French Guiana
  Tel: 594-33-51-11
  Fax: 594-33-47-66
Arianespace Inc.
  700 13th Street, NW, Suite 200
  Washington, D.C. 20005
  U.S.A.
  Attn: Michelle Lyle

India, Srihirakota Range

In the 1990s, India plans to launch approximately 10 Indian spacecraft from the Srihirakota Launch Range (SHAR). The SHAR is capable of test, assembly, and launch of large multistage rockets for Indian spacecraft. The range is located on Srihirakota Island on India's eastern coast approximately 100 km north of Madras. The island is in the Nellore District of Andhra Pradesh State which lies along the east of Pulicat Lake. The SHAR is reached by road from Madras. Prior to its acquisition by the Indian government, the island was a firewood plantation.

The total area of the SHAR is about 145 km^2 with 27 km of coastal length. The range became operational in October 1971 when three Rohini 125 sounding rockets were launched. The facilities at the SHAR have expanded since that time. There is a solid propellant space booster plant which processes large-size propellant grains for the launch vehicle and stage motor programs. Also, there is a static test and evaluation complex which tests for qualification the different types of solid motors used for the Augmented Satellite Launch Vehicle (ASLV) and Polar Satellite Launch Vehicle (PSLV).

The Srihirakota Launch Range consists of three separate assembly and launch areas. Their construction is indicative of the evolutionary growth in India's launch capability.

Launch Site Description

The first launch site was built in the 1970s to accommodate the four-stage SLV-3 (Satellite Launch Vehicle). This site could place satellites up to 40 kg into low Earth orbit. The site is currently deactivated.

The second site was built to accommodate the ASLV. This site can place up to 150 kg into low Earth orbit. The site has a mobile service structure with a clean room for the integration of the satellite and launch vehicle. The ASLV is assembled in the vertical position beginning in the Vehicle Integration Building and completed on the launch pad within the 40-m-tall Mobile Service Structure. Following two unsuccessful launch attempts in March 1987 and July 1988, this site supported a successful ASLV launch in May 1992.

The third site was built to accommodate the PSLV which was commissioned in 1990. The PSLV is capable of placing a 1,000 kg class satellite into a sun synchronous orbit. The PSLV site is based on the concept of the mobile service structure. The 75-m-high Mobile Service Tower provides a payload clean room at the 41-m level. It is equipped to accommodate the liquid second and fourth stages of the PSLV. Satellite preparation and integration is undertaken in the SP-1 clean room facility.

For the 1990s, the Indian government is planning a launch site to accommodate the Geosynchronous Satellite Launch Vehicle (GSLV), which will be capable of placing a 2,500 kg payload into geosynchronous transfer orbit.

Point of Contact

Srihirakota Range
  Andhra Pradesh 524124
  India
  Tel: 91-2001-041-394
  Fax: 91-2001-041-568594

Italy, San Marco Equatorial Range

The San Marco Equatorial Range is a civil, mobile, equatorial launch site in the Indian Ocean, owned by the Italian government. It has operated since the mid-1960s under the direction of the Aerospace Research Center of the University of Rome. The initial purpose of the range was to launch Italy's San Marco 2 satellite into equatorial orbit using a four-stage NASA Scout launch vehicle. The first such launch was performed in April 1967. By mid-1976, eight satellites had been launched: four Italian, one British, and three U.S. The last use of the platform to launch an Italian satellite occurred in March 1988 (San Marco 5 for the University of Rome and NASA) on a Scout G-1. This represented its first firing in 12 years.

The range was used to launch NASA's Small Astronomy Satellite Uhuru (alias Explorer 42) in December 1970. This was the first U.S. satellite put into orbit using a foreign launch base. The advantage of this equatorial site is that the 195 kg Explorer 43 can be launched on a Scout instead of on a larger rocket required for a Cape Canaveral launch. In addition to the Scout facilities, there exists a sounding rocket ramp for vehicles such as the Nike-Apache and Nike-Tomahawk.

In 1993, the Italian space program underwent a process of reorganization and reassessment of future space program objectives. As a result, the San Marco-Scout launcher program, an Italian launcher based on the U.S. Scout vehicle, was suspended. A new indigenously designed vehicle, the Vega, was assessed to be an alternative for Italy's future needs. The Vega is currently in the late design phase and the basic booster, designated Zefiro, has been successfully tested. If this new launcher program is approved, Vega could be available in the 1997–1998 time frame.

Launch Site Description

The San Marco Equatorial Range consists of two distinct offshore platforms (one for launch and one for control) in Ngwana Bay, approximately 5 km off the African coast near Kenya. The San Marco launch platform and the Santa Rita control platform are about 500 km apart and are linked by 23 cables. These two platforms are in radio contact with ground facilities (a main station and a mobile telemetry station) in Kenya.

The San Marco launch pad and the support equipment required to assemble and test Scout-class rockets are floated on a large steel pontoon consisting of 20 steel legs sunk into the seabed and weighing about

3 million kg. The hull is rectangular (30 m by 100 m) and equipped with 20 steel caissons that can be embedded into the ocean floor using a pneumatic system. Electrical power for normal activities is supplied by eight 100 kW diesel generators. However, during final countdown activities, power is provided from the Santa Rita generators by submarine cables. Launch vehicle assembly and testing is performed horizontally in a shelter at the rear of the San Marco launch platform. The shelter is pulled along rails to the launch pad. The launcher and rocket are then pointed to the azimuth to place the payload in the proper equatorial orbit.

The Santa Rita control platform is a three-legged modified oil rig. The triangular structure is about 40 m on each side, and its trestle-like steel legs are operated electrically. There are six 100 kW diesel generators located on a small sea platform. The platform includes the range control center, the vehicle control center, and telemetry stations. The platform can support 80 people.

Points of Contact

Agenzia Spaziale Italiana (ASI)
   250 E Street, SW, Suite 30
   Washington, D.C. 20024
   U.S.A.
   Attn: Mr. Enzo Letico
Agenzia Spaziale Italiano (ASI)
   Via di Villa Patrizi 13
   00161 Rome
   Italy
   Tel: 39-6-440-4205
   Fax: 39-6-440-4212

Japan, Kagoshima Space Center

The Kagoshima Space Center was established in 1962 in the southern part of Japan facing the Pacific to launch sounding rockets and low Earth orbit scientific satellites. The Institute of Space and Aeronautical Science (ISAS) of the University of Tokyo selected a new launch site on several uninhabited hills facing the Pacific Ocean near Uchinoura in the Kagoshima Prefecture. The space center officially opened in December 1963. Since its launch facilities are located on eight level hilltops at different heights covering 0.71 square km^2, Kagoshima looks unique. Kagoshima is arranged on Nagatsubo Plateau about 320 m above sea level.

The Institute of Space and Aeronautical Science of the University of Tokyo has also developed a family of solid rockets and scientific payloads to be launched from the Kagoshima Space Center. By 1965, Kagoshima was equipped to launch the solid fuel Kappa and Lambda sounding rockets. By 1966, Kagoshima was capable of launching a small satellite using a Lambda-class, four-stage rocket.

In February 1970, a Lambda-class launcher successfully placed Japan's first satellite (Ohsumi) into orbit. Since then, the Mu-class family of launchers has undergone continuous upgrading to include the currently operational M-3SII launcher. A new heavier payload M-5 launcher is expected as a follow-on to the M-3SII in the 1995 time period. (The last launch of the M-3SII is expected in 1994.) Scientific satellites have been launched from Kagoshima at a rate of about one per year.

Seasonal launch windows from Kagoshima are restricted to 45-day launch seasons in January/February and August/September because of range safety procedures and complaints from the fishing industry.

Launch Site Description

The Lambda launch complex is about 277 m above sea level. It includes several mobile launch pads and an explosives handling room.

The Mu launch complex is about 220 m above sea level. It is the largest of the Kagoshima Space Center and contains the Mu service tower, assembly shop, launch control blockhouse (about 80 m away from the launch pad), satellite test shop, dynamic-balance test shop, and propellant store. The Mu launch vehicle is assembled on its launcher inside the service tower in a vertical position. Once the vehicle and spacecraft assembly is complete, the service tower is oriented toward the launch direction. The launcher with the Mu launch vehicle is drawn out of the service tower and tilted to the appropriate (75 to 80°) launch elevation angle.

In the 1980s, the Mu launch complex was upgraded to support the more powerful Mu-3SII vehicle. This launch vehicle launched the Japanese Halley's Comet probes Sakigake (Pioneer) and Suisei (Comet) in January 1985 and August 1985, respectively.

Point of Contact

Kagoshima Space Center
    Uchinoura, Kimotsuki-gun
    Kagoshima 893-14
    Japan
    Tel: 81-99-467-2211
    Fax: 81-99-467-3811

Japan, Tanegashima Space Center

From the time Japan started to build an indigenous rocket launch capability in the 1950s, its plan was to gradually develop an independent launch capability for geostationary satellites. Since 1964, the facilities on Niijima Island in the Bay of Tokyo were used to develop sounding rocket technology. To launch satellites, it was necessary to replace the Niijima Test Center by new launch facilities on the southeast tip of Tanegashima Island (100 km to the south of the Kagoshima Space Center and about 1,000 km southwest of Tokyo). The Tanegashima Space Center is now comprised of the Osaki Launch Site (deactivated space launch facility), the Yoshinobu Launch Site (operational H-II facility), and the Takesaki Launch Site (active sounding rocket facility). Space launches from Tanegashima are restricted to 45-day launch seasons in January/February and August/September because of range safety procedures and complaints from the fishing industry.

The Osaki Launch Site was established to operate the N-I vehicle based on the Thor-Delta first stage and a Japanese engine (LE-3) second stage. On February 23, 1967, an N-I vehicle placed the Kiku-2 satellite in geostationary orbit (making Japan, at the time, the only country besides the U.S. and the U.S.S.R. to achieve geosynchronous orbit). The Osaki Launch Site was upgraded to accommodate the improved N-II launch vehicle which launched a total of eight satellites between 1981 and 1987, including Japan's first sun synchronous satellite (the Momo-1 for remote sensing observations).

The National Space Development Agency (NASDA) has done considerable research into cryogenic propulsion for launching heavier satellites for Earth observations and direct TV broadcasts. The goal is to provide Japan with an autonomous heavy lift capability and to become competitive for commercial launch services. The two-stage N-II was replaced by the three-stage H-I vehicle developed by replacing the N-II second stage with a cryogenic (LOX/LH$_2$) stage utilizing an engine (LE-5) with re-ignition capability. The Osaki Launch Site was modified to handle liquid hydrogen. The H-I launch vehicle was used to launch Japanese satellites of approximately 550 kg into geosynchronous orbit and of approximately 1,500 kg into sun synchronous orbits.

The NASDA recently tested a new heavy lift launch H-II, a two-stage cryogenic rocket equipped with two large, solid, strap-on rocket boosters at the Yoshinobu Launch Site (built from 1986 to 1992). The H-II will be capable of launching approximately 2,000 kg into geostationary orbit, 4,000 kg into geosynchronous transfer orbit, and 10,000 kg into low Earth orbit. Its second stage is an improved version of the LE-5 engine, while its first stage employs a newly developed liquid hydrogen/oxygen LE-7 engine. The goal of the H-II program is to develop a heavy lift capability with a lower cost and high reliability.

The Rocket System Corporation (RSC) of Tokyo, Japan, was established in July 1990 as a private organization for marketing the commercial launch services of Japan's new H-II launcher. Working capital was provided by over 75 Japanese industrial and business organizations. The RSC is responsible for H-II procurement, contracting, and management of the commercial launch operations. The commercial H-II will be launched from the Tanegashima Space Center. A new complex called the Yoshinobu Launch Site has been completed for the H-II launches under the direction of the RSC with the technical and facility support of NASDA. The site is located on Tanegashima Island near the Osaki Launch Site.

Launch Site Description

Osaki Launch Site Description

The Osaki Launch Site employs a 6.4-m-high and 12-m-wide launch deck (170,000 kg mass) with a cantilevered steel structure. There are two umbilical masts (43 m and 49 m) that support the feed pipes and conduits for liquid oxygen/hydrogen, electrical power, air conditioning, etc. There is also a mobile service tower (MST) with dimensions of 67 m in height, 26 m in width, and 25 m in depth. The mass of the MST is 3 million kg. Prior to launch, the MST moves approximately 100 m away from the launch pad.

The first two stages of the launch vehicle are processed at Vehicle Assembly Buildings 1 and 2 prior to erection on the pad. There is a Solid Motor Test Building for solid motors and pyrotechnics. A Non-Destructive Test Facility can accommodate ultrasonic and x-ray inspection of upper stages and apogee kick motors. The Spin Test Building accommodates integration of upper stages with its spin table. The Spacecraft Test and Assembly Building processes payloads in a class 100,000 clean room in preparation for mating with the upper stage which is performed in the Third Stage and Spacecraft Assembly Building.

The Osaki Launch Site was upgraded to handle the H-I by adding an additional cryogenic propellant facility and another taller service tower to the existing N1/2 complex. The first H-I test launch was successfully performed in August 1986. The last launch was performed on February 11, 1992. The site is no longer used.

Yoshinobu Launch Site

The Yoshinobu Launch Site was built to accommodate the launching of the larger H-II vehicle from the headland further west. The main facilities

The Tanegashima Space Center is on the southwest tip of Tanegashima Island, about 1,000 km southwest of Tokyo.

of the Yoshinobu Launch Site are a vehicle assembly building, a mobile launcher, and a pad service tower. There are facilities for storage of cryogenic propellants and a blockhouse. To achieve a launch rate higher than two per year, the H-II vehicle will be assembled in the vehicle assembly building and transported to the pad service tower (500 m away) via a mobile platform. There will be two mobile platforms available so that one H-II vehicle can be assembled while a second H-II vehicle is on the launch pad. The first H-II launch took place on February 4, 1994.

The Yoshinobu Launch Site also includes a static test fire stand for the LE-7 engine and a computed tomography unit for nondestructive testing of solid rocket booster segments.

Takesaki Launch Site

The Takesaki Launch Site is located about 2.5 km south of the Osaki Range. The site was operational in 1968 and is comprised of two launch pads. In 1988 through 1989, the site supported launches of the TR-1 single stage rocket that was utilized in development testing in support of Japan's H-II launcher. A new, improved follow-on version, the TR-1A, is currently being offered for commercial microgravity experiment launches.

The H-II Launch Vehicle is Japan's first purely domestic two-stage rocket. It uses liquid hydrogen and liquid oxygen for its first- and second-stage engines.

Point of Contact

Tanegashima Space Center
　Minamitane-machi, Kumage-gun
　Kagoshima 891-37
　Japan
　Tel: 81-99-726-2111
　Fax: 81-99-726-0199

Kazakhstan, Baikonur (Tyuratam) Satellite Launch Center

The Baikonur Cosmodrome lies due north of the railway junction town of Tyuratam east of the Aral Sea about 370 km southwest of Baikonur. It is not located at the town of Baikonur (47.5°N., 65.5°E.). The Republic of Kazakhstan became the owner of the Baikonur Cosmodrome following the dissolution of the Soviet Union. Kazakhstan named the site after Tyuratam in 1992. Kazakhstan does not make direct commercial launch service offerings, but will, by agreement, provide the facilities to commercial users. Several organizations within the Republic of Russia are currently offering a variety of launch services that require the use of the Baikonur facilities. Russia, by agreement, provides the specialist personnel who are needed to perform launch processing and facility operations functions. The Ukraine has also proposed commercial launch services that would utilize the Baikonur facilities.

The Baikonur Cosmodrome is a huge facility in terms of area. That part which is covered by its "Y" configuration is approximately 137 km long and 55 km wide. It is supported by an extensive system of railways and airports. Construction of the first pad was begun in the early 1950s. Both Sputnik 1 and Yuri Gagarin were launched from this pad. This main pad is now designated as the Gargarin Launching Complex and has supported more than 300 Soyuz space missions. A second pad with capability for manned launches was built in 1965, approximately 30 km from the first pad. The site is about 32 km from the "science city" of Leninsk (which itself is 2,100 km from Moscow).

Four launch pads were constructed for the Proton launcher: two in the 1960s and two in the 1970s. Three pads are currently operational; one is being refurbished. The most recent set of pads has supported more than 50 launches per pad.

Launch Site Description

All Soyuz and Progress spacecraft are integrated in the Space Vehicle Assembly-Test Building. The Soyuz-class launcher vehicles are processed separately. The spacecraft proceeds on a special transporter to the fueling facility for loading of noncryogenic propellants and pressurant gases. The spacecraft then returns to the main Space Vehicle Assembly-Test Building where it is mated to the launch vehicle on a rail mounted transporter-erector to undergo final checkout. Due to the extreme climate, all launch preparations are transferred after mating on the rail transporter-erector by diesel locomotive to the main Soyuz pad about 5 km away. Facilities for fueling are located in remote subterranean rooms hundreds of meters from the pad. A shallow, underground command bunker is located near the launch pad.

The Energiya and Shuttle processing areas are to the northwest of the Soyuz buildings. Further to the northwest (about 12 km) is the 4,500-km-long and 84-m-wide Shuttle landing strip on a northeast to southwest orientation. An extensive launch site infrastructure supports the new Energiya heavy launch vehicle and the VKK (Vozdushno-Kosmichesky Korabl, or Air-Space Vehicle). There are three Energiya launch pads. One pad was specially constructed for testing and launching the basic Energiya vehicle. A set of two pads for the Energiya/VKK were converted from the mothballed facilities associated with the previous Soviet Type G lunar rocket (N-1) program. The single pad is used for launches of conventional side-mounted payloads. The set of two pads was designed to support launches of the VKK. (At least one or possibly both of these pads are associated with the new Energiya-M program.) Each of the launch pads is connected by a double railway track to an assembly and checkout building that is used for assembly of the large Energiya booster components.

The Proton preparation and launch area is west of the Shuttle landing strip many kilometers from the main Baikonur base. At least two launchers can be handled simultaneously in this large horizontal building. The Proton is transported to the launch pad fully assembled with the spacecraft in position ready for launching. For commercial customers, payload facilities are available about 70 km from the Proton integration buildings. There is a 270 m^2, class 100,000, temperature-controlled Satellite Preparation Room with an adjacent 100 m^2, class 10,000 clean room. The payload can then be transferred into a separate 100 m^2, temperature- and cleanliness-controlled hydrazine and nitrogen loading area. Next, the temperature-controlled, 500 m^2 General Assembly Room is used to accomplish the horizontal attachment to the spacecraft adapter, final stage, and fairing. Other subsystem testing and storage facilities also are available.

Point of Contact

The Space Commerce Corporation
 5718 Westheimer, Suite 1515
 Houston, TX 77057
 U.S.A.
 Tel: 713-977-1543

Norway, Andoya Rocket Range

From the start of operation in 1962, more than 400 sounding rockets have been launched from the Andoya Rocket Range. The size of these vehicles has ranged from the Dart to the Black Brant X. In addition, over 159 scientific balloons have been launched from the range. The range is supported by the European Space Agency and is available to member nations on a marginal cost basis. The range is available to other customers on a full cost basis. Currently, the range has a support staff of about 25.

The Norwegian Space Center (NSC) and the Swedish Space Corporation are currently studying the design of a dedicated launch facility that could take advantage of existing facilities and geographical location to offer integration, launch, and operation of small satellites on a commercial basis. The facility will be based on the existing infrastructure of the Andoya Rocket Range in Norway and the Esrange, the operations center of the Swedish Space Corporation. Launches will occur from a new site close to the current Andoya Rocket Range, on the island of Andoya in northern Norway.

Technical, operational, and market aspects of supplying and providing a commercial launch service for small satellites was studied in a joint Norwegian-Swedish project in 1992 and 1993. If development starts in 1994 as planned, the first launch of small satellites into polar orbits is expected in 1996

The small satellite launch service is limited to launch vehicles with less than 40,000 kg of net explosives due to ground safety considerations and proximity to inhabited areas. This limitation on launch vehicle mass constrains the orbital capacity to 225 kg into a 750 km sun synchronous (98°) circular orbit. Different launch vehicles which could provide a low cost service have been assessed in feasibility studies conducted by the range. The Russian START-1 and Pac-Astro's PA-2 launch vehicles have been found to be most compliant with the requirements.

Launch Site Description

There are currently eight launch pads at Andoya Rocket Range. The largest two pads have a 3,300-kg load limit. Upgrades that would allow launch of vehicles up to 20,000 kg have begun. Also, work has begun on a new launch ramp and a more accurate trajectory and position tracking system. There is access to a NATO runway (5,000 m) close by.

The major new infrastructure at the Andoya Rocket Range required to upgrade the facility to launch small satellites is a new launch pad and a blockhouse facility. The assembly tower will be based on the MAXUS tower design recently constructed at Esrange. A new payload integration and test facility is also planned. However, this facility will not be constructed unless vertical integration is required.

The joint Norwegian-Swedish study also found that a turnkey service, covering the launch vehicle together with its launch and flight operations, is most appropriate for providing a minimum cost service to potential customers.

Point of Contact

Norsk Romsenter (Norwegian Space Center)
  Hoffsveien 65A
  P.O. Box 85 Smestad
  N-0309 Oslo 9
  Norway
  Tel: 47-2-52-38-00
  Fax: 47-2-23-97

Pakistan, SUPARCO Flight Test Range

The Space and Upper Atmosphere Research Commission (SUPARCO) is Pakistan's national space agency and operator of the national sounding rocket Flight Test Facility. Although orbital launch operations are not available, sounding rocket services for scientific research could be made available through negotiations.

A Flight Test Range is located about 50 km northwest of Karachi at Miani Beach, Sonmiani, in the district of Labella. It covers an area of about 500 acres. The range was established in the early 1960s for scientific research of the upper atmosphere. Current plans include the future upgrade of the range to accommodate larger-class sounding rockets. The range is operated by the SUPARCO.

Typical sounding rockets are the Centaure and the Dragon (French origin, Pak-Shahpar). Typical capability is lift of a 60 kg payload to an altitude of 440 km. The launch corridor azimuths are limited from 220 to 310° (sea to Sonomi Bay). The preferred launch season is during the clear weather season during October to March.

The SUPARCO Flight Test Range provides sounding rocket services for scientific research.

Launch Site Description

Fixed rail launchers of various types exist on separate launch pads. These are supported by a mobile crane and mobile service platforms. Facilities at the SUPARCO Flight Test Range include an assembly shop (rocket and payload integration hall), an electronics maintenance shop, a 60-m wind tower, and other meterological support equipment.

Port facilities are available in Karachi (50 km to the southeast of the range).

Payload Accommodations

Not available at SUPARCO Flight Test Range.

Point of Contact

Pakistan Space and Upper Atmosphere Research
  Commission (SUPARCO)
  Sector 28, Gulzar-e-Hijri
  Off University Road, P.O. Box 8402
  Karachi 75270
  Pakistan
  Attn: M. Nasim Shah
  Tel: 474261-4
  Fax: 92-21-496-0553

Russia, Plesetsk (Northern) Cosmodrome

Plesetsk Cosmodrome is located 170 km south of Archangel at a latitude which enables satellites to be placed in polar and highly elliptical orbits. Its existence was established following its first orbital launch (Cosmos 112) in March 1966; however, its existence was not officially acknowledged until 1983. The Plesetsk Cosmodrome (also called Mirny) is known to Russians as the Northern Cosmodrome and has similarities to Vandenberg Air Force Base in the United States. Both contain a large number of launch pads as well as underground silos, and both are responsible for launching payloads into high inclination orbits. Since its establishment in the 1960s, Plesetsk has been the world's busiest space center, supporting over 1,200 launches (more than the United States, Europe, Japan, and China combined). Though near to the town of Plesetsk, the cosmodrome is colocated with the town of Mirny, a name often applied to the cosmosdrome.

Plesetsk has been used as the launch site for a wide variety of photo-reconnaissance, communications, navigation, early warning, scientific, and other satellites. The highly elliptical Molniya satellites are also launched from the Plesetsk site.

It is expected that activity at the Plesetsk Cosmodrome will substantially decrease in the 1990s. When the Glonass high-altitude navigation network (launched from Baikonur Cosmodrome) is operational, the launches of the smaller, low-altitude navigation satellites (about nine per year) may be phased out. Also, the use of highly elliptical early warning satellites may be phased out in favor of a new constellation of geostationary satellites. Finally, the use of 2-week photo reconnaissance satellites (typically launched from Plesetsk) may be significantly decreased in the 1990s.

Launch Site Description

As of 1991, there were thought to be nine operational pads (though some may be used for missile testing). Four launch pads are capable of supporting launches of the most powerful vehicle at Plesetsk, the Soyuz-class launcher. Two pads are dedicated to the Tsyclon launcher that was initially launched in 1977. The Tsyclon is processed in a horizontal manner in a large, rail-served, six-storied building. Launch facilities for the Cosmos launcher are also available. A Zenit launch site, along with necessary launch vehicle and spacecraft processing facilities, is currently under construction to accommodate future heavy-lift polar and low Earth orbit needs. Construction of a Proton launch facility is also being considered. All new construction is pending political developments.

In March 1993, a new small-class satellite launcher, the START-1, was test launched from Plesetsk. The launcher utilized stages derived from the former Soviet SS-25 ICBM and demonstrated the peaceful use of ballistic missile technology for space applications.

Points of Contact

The Space Commerce Corporation
   5718 Westheimer, Suite 1515
   Houston, TX 77057
   U.S.A.
   Tel: 713-977-1543
Glavkosmos
   9, Krasnoproletarskaya St
   Moscow 103030
   Russia
   Attn: Alexander Dunaev
   Tel: 972-44-97
   Fax: 288-95-83

United States, Cape San Blas
Launch Operations

Cape San Blas launch facility is located on the west coast of Florida at approximately 29°43'N., 85°20'W. The location is on the panhandle of Florida near the Eglin Air Force Base range. The launch site is situated on the Gulf of Mexico approximately 100 miles southwest of Tallahassee and 15 miles south of Port Saint Joe.

During the 1950s and 1960s, the U.S. Air Force undertook an extensive sounding rocket test program from Florida's Panhandle. Cape San Blas was a center for this activity. Test Site D-3A included three active launch pads from which a variety of rockets were launched over the Gulf of Mexico. While no Air Force launches have been conducted at the site for decades, D-3A is still an active military facility where radar tracking and other operations are routinely performed.

The Cape San Blas launch facility was reactivated by the Spaceport Florida Authority (SFA) as a suborbital launch facility for timely, low cost access to space for university micro payload experiments. Due to environmental sensitivity, the activities at Cape San Blas are limited. Only sub-orbital rockets may be launched from Cape San Blas and construction of support facilities must be at least 5 miles from the launch site. First launch from the reactivated Cape San Blas site of an upper atmospheric meteorological sounder for Florida State University's Meteorological Department occurred in 1992.

The Microstar launch system is manufactured by Orbital Sciences Corporation's Space Data Division (SDD) in Chandler, Arizona. The Microstar system was originally developed for the U.S. Navy as a surface-to-air missile. These systems have also been used by various governments since 1968 under the Cooperative Meteorological Rocket Network to obtain information about the upper atmosphere. SDD manufactures several variations of the Microstar vehicle. Over 20,000 Microstar launches have been conducted with no known major failures.

Stringent launch criteria have been established for the Microstar launch system at Cape San Blas by the Spaceport Authority. Launches are conducted only in daylight hours with at least 8 miles of horizontal visibility. Launches will not be conducted under significant cloud cover or in winds exceeding 49 knots. Prior to each launch, mariners and aviators will be notified and beach access south of the launch site will be closed.

The first launch from the reactivated Cape San Blas site occurred in 1992.

Launch Site Description

Equipment needed to launch the Microstar rockets from Cape San Blas includes a launch rail, a GMD-5 transportable telemetry/tracking system, and a mobile launch control van. The launch rail, bolted onto one of the existing concrete launch pads (which are approximately 15 feet by 15 feet), is the only permanent fixture at the site.

The Microstar helical rail assembly provides a spin-stabilization of 16 revolutions per second. The launch rail assembly is aimed to target a nominal ballistic impact area approximately 3.5 nautical miles to the south downrange in the Gulf of Mexico. Payload recovery is an optional capability.

Point of Contact

Spaceport Florida Authority
    Cocoa Beach, FL
    U.S.A.
    Attn: Edward Ellegood
    Tel: 407-868-6983

United States, Eastern Range

The Eastern Range (ER) launch facility is located at Cape Canaveral Air Station (CCAS) on the Atlantic Coast of Florida at approximately 28°28'N., 80°34'W., about 69 miles east of Orlando, Florida. It is situated on a barrier island between the Banana River and the Atlantic Ocean along the central Florida coast. The CCAS extends along an expanse of coast and provides overwater flight for low inclination and equatorial orbits. The nearest major communities are Cocoa Beach and Cape Canaveral. ER headquarters are located at Patrick Air Force Base, which is 21 miles south of CCAS. The CCAS location between the Atlantic Ocean and Banana River provides a natural safety buffer around the launch area. The range also includes instrumentation sites at island stations as well as along the Florida coast that can support suborbital launches as well as mobile air and sea launches from the Atlantic Ocean area.

The installation at Patrick Air Force Base was established by the U.S. Navy during World War II. Activated in 1940 as the Banana River Naval Air Station, it served as a base for antisubmarine sea-patrol planes during the war. In 1948, it was transferred to the Air Force and maintained in a standby status awaiting activation of the Joint Long Range Proving Ground (JLRPG). Later the station was renamed JLRPG Base and in 1950 it received its present name, Patrick Air Force Base. Responsibility for developing and operating the range was given to the Air Force in May 1950. The first missile launch, a German V-2 rocket with an Army WAC-Corporal second stage, occurred in 1950.

CCAS encompasses 15,800 acres and occupies a barrier island composed of relic beach ridges formed by wind and wave action. Approximately 70 percent of CCAS has been retained in a natural state of either virgin stand or secondary growth of vegetation. The island parallels the mainland separating the Atlantic Ocean from the Banana River and the Indian River Lagoon. The island is approximately 4.5 miles (7.3 km) wide at its widest point. The land surface elevation ranges from sea level to 20 feet above mean sea level. The station has 14.5 miles (23.2 km) of ocean coastline and 12.2 miles (19.5 km) of river shoreline.

The first U.S. orbiting satellite was launched from CCAS in early 1958. Since then, more than 400 orbital and suborbital space launches have been conducted here. In addition, more than 70 space launches have occurred from nearby Kennedy Space Center. CCAS includes a Navy Trident Launch Facility, space launch complexes, and an aircraft landing strip. Instrumentation

Located on a barrier island, Cape Canaveral Air Station offers six active fixed launch complexes for commercial and government missions.

and communication sites are distributed throughout the base as well as at several offsite locations both up-range along the Florida coast and down-range in the Atlantic. Space launches have the option of acquiring telemetry using Advanced Range Instrumentation Aircraft. In addition, the ER may use other DOD or NASA facilities at Argentia, Newfoundland; Wallops Flight Facility, Virginia; and Bermuda. Launch azimuths for orbital launches are allowed between 37 and 112°.

There are presently six active fixed launch complexes on CCAS from which space launches are conducted. These are Space Launch Complexes (SLC) 17A and B (Delta), 36A and B (Atlas), 40 (Titan III and IV), and 41 (Titan IV). In addition, air-dropped Pegasus space launches are performed offshore. Ballistic launches are conducted from pads SLC-20 or SLC-46 into the Atlantic broad ocean area.

Both orbital and suborbital commercial space launches can be conducted from the Eastern Range as well as launches by the U.S. Government. Currently, commercial space launches are available from Delta II, Atlas II, Titan III, and Joust. Air-launched Pegasus missions which are conducted offshore over the Atlantic originate from the Pegasus operations facilities at Wallops Flight Facility or Vandenberg AFB. Future capabilities are planned for several new launch vehicles, including the Lockheed Launch Vehicle (LLV) and Taurus. Under the state chartered Spaceport Florida Authority (SFA), the new commercial space launch developments are expected to be multi-use facilities as opposed to the dedicated, uniquely operated complexes belonging to the U.S. Government. A new spaceport is being developed at SLC-46 to accommodate these new launchers.

The central Florida coast has a subtropical climate. Lightning is a major concern as the cape region has a high frequency of thunderstorms and lightning occurrences. Hurricanes usually skirt around CCAS and strike with reduced wind force when they do occur.

Launch Site Description

Atlas Launch Operations

The first Atlas space launch occurred in 1958. Since then, the Atlas vehicle has been used in more than 144 space launches from CCAS, including planetary, lunar, and Mercury manned spacecraft. The current family of launch vehicles use the Atlas/Centaur configuration. Modifications to Atlas launch facilities to accommodate the latest solid motor thrust augmented vehicle configuration were recently completed. First launch of the Atlas II AS vehicle from the Eastern Range occurred in December 1993. The Atlas CCAS launch site can accommodate the Atlas I, II, IIA, and IIAS configurations.

The Atlas launch facility is located at CCAS Space Launch Complex (SLC) 36. The major facilities include two launch pads, 36A (used for U.S. Government and commercial launches), and 36B (commercial launches only), a common blockhouse, servicing, and support facilities. The launch pads on SLC-36A and SLC-36B include a mobile service tower, fixed umbilical tower, and a launch service building. Vehicle integration is performed on the pad and includes installation of the Atlas booster, mating of the Centaur upper stage, mating of strap-on solid motors (36B only), and mating of the encapsulated spacecraft. Mission operations and data display are at the Mission Director's Center (MDC) in Building AE. For launch operations, the Mission Director and senior management personnel are in the MDC. They are backed by engineering personnel in the data station and the Launch Vehicle Data Center (LVDC) in Building AE.

The mobile service tower is an open steel structure with an interior enclosure that contains retractable servicing and checkout platforms, launch vehicle electrical and mechanical equipment, payload interface equipment, and vehicle erection equipment. The launch service building includes interconnecting ground equipment and also serves as the launch platform. The blockhouse serves as the operations and communications center for the launch complex. The blockhouse also includes launch control, instrumentation and computer equipment to perform launch operations from either SLC-36A or SLC-36B. Vehicle transfer from airship transporters to ground handling trailers and storage is performed at hangars in the CCAS industrial area.

Payload processing is performed offpad at several processing sites both on and off CCAS proper. These include both nonhazardous and hazardous processing facilities. Spacecraft preparation and checkout activities occur in nonhazardous facilities. Hazardous facility operations include high pressure tests, solid motor installation, fueling, ordnance installation, and balance. Following hazardous operations, the spacecraft is encapsulated in the Atlas payload fairing at the payload processing facility, transported to SLC-36A or SLC-36B, and mated with the launch vehicle. The remote payload encapsulation approach is the baseline for Atlas launch vehicle integration. This approach has been used successfully for all Atlas/Centaur operations.

Delta Launch Operations

The Delta launch vehicle was first launched from the ER in 1960 by NASA. Since then, there have been 185 launches from CCAS by the Delta program. The Delta launch site can accommodate the Delta II 6925 and 7925 three-stage configurations as well as 6920 and 7920 two-stage vehicles. The 7925/7920 vehicles have the more powerful strap-on solid motor configurations.

The SLC-17 complex consists of two launch pads (SLC-17A and 17B), a blockhouse, operations buildings, horizontal processing building, and other facilities necessary to prepare, service, and launch the Delta vehicle. Each launch pad includes a mobile service gantry and fixed umbilical tower for access/interface to the Delta launch vehicle and spacecraft. The Delta II launch vehicle is fully integrated on the launch pad. Upper level enclosure, or white room areas, provide protection for the spacecraft, checkout equipment, and personnel while on the launch pad. Launch operations are controlled from the blockhouse which is equipped with vehicle and ground systems monitoring and control equipment. The Mission Director and senior management personnel are in the MDC. They are backed by engineering personnel in the data station and the Launch Vehicle Data Center (LVDC) in Building AE. Mission operations and data display are at the Mission Director's Center in Building AE in the CCAS Industrial Area.

All Delta launch vehicle components, except for solid motors, are processed through checkout facilities in the CCAS Industrial Area. These include first and second stages, payload fairings, and interstages. Ordnance installation and erection preps are performed at the hazardous processing facilities (HPFs) near SLC-17. A solid motor building is also located near SLC-17 for processing of the strap-on solid rocket motors. The facility has been modified to handle both the Castor IV A and the larger graphite-epoxy motors (GEM). The facility is used for receiving, inspection, build-up, and preparation for erection and transportation to SLC-17. The solid motor upper stages are prepared in dedicated upper stage facilities or are processed through an available HPF. Payload SRMs can undergo preliminary cold soak, x-ray, and preparation procedures in the ordnance processing areas at CCAS before final assembly and checkout. The primary payload processing facility for Delta commercial spacecraft is the Astrotech facility. Additional spacecraft checkout facilities are available in NASA facilities in the CCAS Industrial Area. These facilities are government-owned, and are shared by other launch operators on CCAS.

Titan Launch Operations

The Titan III commercial launch vehicle configurations are part of the Titan family of launch vehicles which began with the Titan II missiles. The first Titan space launch occurred in 1964 from CCAS. Since then, the Titan vehicle has been used in 68 space launches from CCAS, including planetary and Gemini manned spacecraft. With its large, segmented, solid motor, the Titan III provides a heavy lift capability to LEO or GTO which allows for multiple payload manifesting and a variety of upper stage options. The commercial Titan III operates from SLC-40 which is configured to also accommodate the larger Titan IV operated by the U.S. Government. The Titan IV is also launched from nearby SLC-41. There have been four commercial Titan launches to date; however, there are currently no future flights scheduled. Titan vehicle processing and integration is accomplished offpad at the Titan Integrate, Transfer, and Launch (ITL) Facility. This facility eliminates the necessity for vehicles to occupy the launch pad for extended periods before launch. The major elements of the ITL complex are the Vertical Integration Building, the Solid Motor Assembly Building, the Solid Motor Assembly and Readiness Facility, the Launch Operations Control Center, SLC-40 and SLC-41.

The Vertical Integration Building (VIB) area provides three integration cells for core vehicle integration and checkout, one integration cell for Centaur upper stage mating with base payload fairing and upper stage checkout, and associated areas for vehicle and component receipt, inspection, modification, and erection. The VIB annex and Payload Fairing Cleaning Building are used for payload fairing processing. Also housed in the VIB is the inertial upper stage (IUS) checkout station, used for IUS checkout and launch countdown control. The Titan IV control center is located remotely at the Launch Operations Control Center (LOCC), which is used for all integrated vehicle testing and launch operations. Titan III launch control was previously performed from the VIB. Titan III control functions have yet to be incorporated into the LOCC.

After the core vehicle is assembled and tested, the vehicle is transported to the Solid Motor Assembly Building where the segmented solid rocket motors are assembled, verified, and mated with the core vehicle. The assembled vehicle is then moved on its transporter by rail to SLC-40. The launch complex consists of a concrete pad with fixed foundations supporting the launch vehicle transporter, the umbilical tower, equipment building, mobile service tower with the payload environmental shelter, and propellant and gas storage areas. At SLC-40, the entire vehicle undergoes a combined systems test using interface simulators to verify launch vehicle readiness before payload mate. After the mobile service tower is moved into position around the vehicle, the aft payload carrier with the encapsulated payload is hoisted from the transporter to the environmental shelter, and then mated to the Titan III. Subsequently, the forward payload carrier is hoisted and mated on top of the aft carrier. Time is allocated for each payload to perform standalone inspections and testing immediately after mating. Combined tests with the Titan III are completed before continuing with final vehicle preparations and launch.

Commercial and government facilities are available for processing payloads to a flight-ready condition. These facilities are shared among customers of other launch vehicle systems. These include both nonhazardous and hazardous processing facilities (HPF). Upper stages are prepared in dedicated upper stage facilities or are processed through an available HPF that can accommodate the vehicle size and weight. Payload SRMs can undergo preliminary cold soak, x-ray, and preparation procedures in the ordnance processing areas at CCAS before final assembly and checkout. Payload encapsulation within the payload carriers occurs at either the nonhazardous processing facility or a specially designated payload encapsulation facility, depending on the payload's physical size and processing requirements. The payload and carrier are assembled and loaded onto the transporter and subsequently moved to SLC-40.

Payload Checkout Facilities

Building AE

Building AE is a Butler-type structure which is completely environmentally controlled. This building contains a spacecraft laboratory and a class 50,000 high bay clean room complex. Also located in the building are the Mission Director's Center (MDC) and offices for the spacecraft contractor and spacecraft management personnel. This building also houses the communications equipment that links the Astrotech facility with NASA and USAF voice and data networks to KSC and CCAS.

Building AM

Building AM is adjacent to Building AE and was built specifically for spacecraft prelaunch operations and checkout. Two floors of offices, conference rooms, a clean room complex, and test areas adjoin the high bay, which contains two spacecraft laboratories. Air conditioning in the building is for personnel comfort only.

Building AO

Building AO has a large clean room for nonhazardous spacecraft processing only. This facility can support two spacecraft projects simultaneously. When supporting two projects, it is functionally divided into two sections. The airlock is the dividing line in the high bay area, while the lobbies constitute the first and second floor dividing lines. The operational areas of this facility consist of a class 100,000 high bay clean room, an airlock, two spacecraft system test areas, one operational center, laboratories, storage areas, and offices for project personnel.

Hazardous Processing Facilities

Explosive Safe Area (ESA) 60A

ESA 60A is located approximately 1 mile (1.6 km) northwest of the CCAS Industrial Area. Major activities performed in this facility are liquid propellant transfer tests and loading operations, ordnance installation, solid propellant motor installation, pressurization tests, and spacecraft assembly. Two interfacility communication systems tie ESA 60A with Buildings AE, AM, and AO. Closed-circuit television cameras in the Spacecraft Assembly Building and Propellant Laboratory can be connected to one or more monitoring stations in Buildings AE, AM, and AO. Voice and data links can also be established with the SLCs and the Astrotech facility. The Spacecraft Assembly Building is a high bay structure containing three operational areas. Two are spacecraft assembly high bays and the other is an airlock. All three are class 100,000 clean rooms. The two high bays are configured identically. The Propellant Laboratory is a two-level structure with earth revetments on the north and south walls. All operations are conducted in the class 100,000 clean high bay. This building has a dynamic balancing machine with a 12,000-lb (5,445 kg) spin test capacity. ESA 60A also houses an instrumentation laboratory and a small cleaning laboratory.

Spacecraft Assembly and Encapsulation Facility No. 2 (SAEF 2)

This is a high-bay HPF located southeast of the KSC Industrial Area on nearby Merritt Island. It is used for assembly, testing, propellant loading, pressurization, and handling of explosive items for spacecraft. The facility offers an airlock, a high bay, two low bays, and a test cell for support of spacecraft operations. Two remote payload control rooms and supporting areas are available. These rooms are located in separate buildings.

NASA Payload Processing Facilities

NASA operates spacecraft checkout facilities in support of nonhazardous payload processing. These facilities are located in the Cape Canaveral Industrial Area in Buildings AE, AM, and AO. Checkout facilities are complemented by hazardous processing facilities for performing liquid propellant, solid motor, explosive device assembly, and dynamic balancing. These facilities include Explosive Safe Area 60 at CCAS, and Spacecraft Assembly and Encapsulation Facility No. 2 at KSC. In addition to these facilities, several other shared facilities or work areas are available for supporting spacecraft projects and the spacecraft contractors at CCAS. These areas include: solid propellant storage area, explosive storage magazines, Missile Research Test Building, nondestructive testing laboratory, Electromechanical Test Facility, and liquid propellant storage area.

Commercial Payload Processing Facilities

The Astrotech commercial payload processing facility is located in the Gateway Industrial Center in Titusville, Florida, approximately 15 miles (24 km) from CCAS. This facility includes 100,000 ft^2 of industrial space constructed on 40 acres of land. The facility is owned by Astrotech Space Operations. There are five major buildings on the site: the nonhazardous and hazardous processing facilities, the environmental and warehouse storage facilities, and the owner/operator office area.

The Nonhazardous Processing Facility is used for final assembly and checkout of the spacecraft. It houses three spacecraft class 100,000 clean room high bays, control rooms, and offices. Antennas mounted on the building provide communication with the SLCs and Building AE at CCAS.

The Hazardous Processing Facility houses three explosion-proof high bays for hazardous operations, spin balancing, payload/solid motor final assembly, and fueling. The dynamic balance machine has spin test capacity of 18,500 lb (7,562 kg).

The Environmental Storage Facility provides six secure, air-conditioned, masonry-constructed bays for storage of high-value hardware or hazardous materials.

The Warehouse Storage Facility provides space for covered storage of shipping containers, hoisting and handling equipment, and other articles not requiring environmental control.

The owner/operator office area is an executive office building that provides the spacecraft project officials with office space for conducting business during their stay at Astrotech and the Eastern Range.

Spaceport Florida Launch Facilities

The Spaceport Florida Authority (SFA) which currently operates a sub-orbital space launch facility at Cape San Blas, Florida, is planning to develop a multi-use commercial launch capability at SLC-46 on CCAS.

The CCAS launch facility is planned for completion and first use in 1995. The SLC-46 site was used for Trident test launches. It will be converted to a multi-use launch facility capable of accommodating a variety of small launch vehicles, including the LLV and Taurus. The launch vehicles and payload will be integrated on the launch pad. Launch control will be performed from the SLC-46 blockhouse.

Point of Contact

45th Space Wing/XP
  Patrick Air Force Base, FL
  U.S.A.
  Tel: 407-494-5938
  Fax: 407-494-7302

United States, Poker Flat Research Range

The Poker Flat Research Range is owned by the Geophysical Institute of the University of Alaska, Fairbanks, and is the only university-owned spaceport in the world. The sounding rocket facility mission offerings are primarily auroral and middle-upper atmosphere research. The site is located approximately 30 miles northeast of Fairbanks and is the highest latitude launch site in the United States. The National Aeronautics and Space Administration (NASA), Defense Nuclear Agency (DNA), Phillips Lab, National Science Foundation, and National Oceanic and Atmospheric Adminstration (NOAA) all provide funding for Poker Flat Research Range through their contracts. NASA is currently funding the annual operating cost of about $1.5 million.

There have been up to 10 major sounding rocket launches (plus a number of meteorological rockets) annually since the construction of the base started in 1968. Well over 200 major sounding rocket experiments have been launched with the Nike-Tomahawk comprising more than 60 of these launches. The Poker Flat Research Range is located adjacent to the Steese Highway. A number of Federal and state agencies, including the Bureau of Land Management and the State of Alaska, are involved in authorizing the impact of sounding rockets and payloads on the 25 million acres of downrange state land. The rocket flight zones are approved by the Federal Aviation Administration which also coordinates air space use during sounding rocket launches. Traffic on the Steese Highway is stopped during rocket launches from the Poker Flat Research Range.

A major upgrade to the Poker Flat Research Range facilities is planned using U.S. Government funds. During the next few years, plans include a new range office complex, a launch command decision center, a lidar research facility, and a new rocket motor storage facility. Also, in July 1991, the State of Alaska commissioned the Alaska Aerospace Development Spaceport Authority to perform a feasibility study of the international market potential for commercial launch activities (small payload, polar orbit flights) from the high latitude Poker Flat Research Range site.

Launch Site Description

The Poker Flat Research Range consists of a 21 km^2 central complex, which includes telemetry hardware, an optical observatory, five active pads, a rocket storage and assembly building, maintenance and communications facilities, a payload assembly building, and a range office. The value of the range facilities is more than $15 million. There is a small number of full-time university employees who work for the range maintaining the facilities and providing launch services, including the applications for the necessary approvals and waivers necessary for each flight. As necessary, employees are added to support individual launches.

There are five principal pads at the Poker Flat Research Range. Both Pad 1 and Pad 2 have a 7.5K launcher while both Pad 3 and Pad 4 have 20K MRL launchers installed. There is a 4K3 twinboom launcher at Pad 5. Pad 6 which had been used for meteorological missions has been deactivated. Missions are typically supported by a ground-based network of stations in Alaska and Canada. The optical observatory includes magnetometers, riometers, all-sky auroral cameras, meridian scanning photometers, and equipment for observing and recording auroral research data. An automated Umkehr measurement station is part of a six-station worldwide network, the Automated Dobson Network (ADN), used to observe and measure ozone. A Stratospheric/Tropospheric radar used to measure turbulence at the tropospheric boundary also is located at the range.

Telemetry at the Poker Flat Research Range is provided by 8- and 16-foot dishes operating in the S-band and provided by NASA. These dishes are located in a 50-foot dome at the top of the hill overlooking the range. In addition, there is a NASA owned and operated mobile C-band radar which was installed at Poker Flat Research Range in 1983. This mobile radar replaced the S-band Verlort radar installed in 1970 and finally deactivated in 1984.

Point of Contact

Geophysical Institute
  University of Alaska, Fairbanks
  Fairbanks, AK 99775-0800
  U.S.A.
  Tel: 907-474-7413
  Fax: 907-474-7015

United States, Wallops Flight Facility

W allops Flight Facility is located on Virginia's Eastern Shore at approximately 37°54' N., 75°30' W., about 40 miles southeast of Salisbury, Maryland, and 150 miles southeast of Greenbelt, Maryland. Wallops consists of three separate areas: the main base, the Wallops Island launch site, and the Wallops mainland.

Wallops Island extends into the Atlantic Ocean on the Delmarva Peninsula and is located about 2 miles from the Wallops mainland facilities separated by an inland waterway connecting Bogues Bay and the Atlantic. The Wallops Flight Facility range includes instrumentation sites on both the mainland and on Wallops Island.

In 1945, the National Advisory Committee for Aeronautics (NACA) added a new dimension to its capability for high speed aerodynamic research when it authorized the Langley Research Center to proceed with the

The Wallops Flight Facility is at the center of NASA's suborbital program, which includes use of rockets, balloons, and aircraft.

development of Wallops Island as a site for research in supersonic flow phenomena with rocket-propelled models. Starting with initial operations in 1945, Wallops Flight Facility was used as a launching site for science and research purposes. For the most part, all requirements have been met with relatively small solid sounding rockets staged in various ways. The largest of the launch vehicles has been the Scout four-stage solid-fuel vehicle, capable of launching small satellites, space probes, and re-entry missions. In 1958, NASA Wallops expanded and took over the Navy Chincoteague Naval Air Station located about 7 miles northwest of Wallops Island which included buildings, utilities, hangars, and an airport.

Today, the facility is at the center of NASA's suborbital programs. Sounding rockets, balloons, and aircraft are used actively in NASA programs concerned with space science, applications, advanced technology, and aeronautical research. Missions are conducted locally and throughout the free world. Twenty-two satellites have been launched from Wallops since 1961, including the first satellite to be launched by an all solid-fuel rocket vehicle.

The Wallops Flight Facility comprises a total of 6,200 acres and includes a research airport and aircraft fleet that are located on the main base. Rocket launch facilities are located on Wallops Island. Processing facilities and control centers are located on both Wallops Island and the mainland. Additional instrumentation belonging to NASA at Bermuda and DOD at the Eastern Range are used to support orbital launches.

There are presently three fixed launch sites on Wallops Island from which space launches are performed: Launch Areas 2 (sounding rockets) and 3 (Scout) and the Conestoga Launch Area. In addition, air-dropped Pegasus space launches are performed offshore. Both orbital and suborbital commercial space launches can be conducted from the Wallops Flight Facility as well as launches by the U.S. Government. Currently, commercial space launches are available from Conestoga and Pegasus. Additional future capabilities will result from the plan to convert the Launch Area 32 Scout facilities for use by the Orbital Express launch vehicle. One blockhouse supports all launch areas.

Launch Site Description

Conestoga Launch Operations

First launch of the Conestoga launch vehicle from Wallops Flight Facility is planned for 1994 in support of the COMET program. The Conestoga launch vehicle family consists of a core solid rocket booster and upper stage with a variety of strap-on solid motor configurations. EER Systems has constructed a new launch facility on Wallops Island near Launch Area 2. Prelaunch processing of launch vehicle and payload also occurs at facilities on Wallops Island.

The Conestoga launch vehicle is fully integrated on the launch pad. Access is provided by a portable service tower. Vehicle processing is performed in the Horizontal Operations Facility, where motor processing, upper stage buildup, ordnance test, and payload/adapter assembly and encapsulation are performed. The facility was previously used as the Scout assembly building. The encapsulated payload is mated to the Conestoga at the launch pad where final integrated testing and servicing take place prior to launch. The launch is controlled from a remote blockhouse modified to support the Conestoga. The mission control/range control functions are performed at the Integrated Control Center on the main base.

Payload processing is performed at existing NASA payload checkout and hazardous processing facilities on Wallops Island.

Pegasus Launch Operations

First launch of the three-stage, air-launched Pegasus launch vehicle occurred in 1993. This flight was ferried from Edwards Air Force Base, California, after processing and integration. Orbital Sciences Corporation, builder and operator of the Pegasus, plans to construct a vehicle assembly building (VAB) at Wallops Flight Facility to support operations from the East Coast. This facility will be similar to the one currently in construction at Vandenberg Air Force Base and will support the new extended-length configuration as well.

All major vehicle subassemblies will be delivered from the factory to the VAB. The Pegasus stages will be integrated horizontally prior to arrival of the payload. The VAB has the capacity to build multiple Pegasus launch vehicles simultaneously. A Pegasus payload is delivered to the integration site 7 to 21 days prior to launch. The payload must complete its own independent verification and checkout prior to beginning integrated processing with Pegasus at the VAB.

Following closeout, the Pegasus and payload are transported to the carrier aircraft for mating at the flight line. Preflight checks begin 4.5 hours before take off. Flight and Launch Control functions are managed from a launch operations control room on the ground. Launch release functions are controlled from onboard the aircraft. The Pegasus is launched over the Atlantic Ocean. The range of achievable inclinations from Wallops Flight Facility are 25 to 65°.

Payload processing is performed at existing NASA payload checkout and hazardous processing facilities on Wallops Island.

Orbex Launch Operations

CTA Launch Services plans to operate its Orbex launch vehicle using excess Scout processing and launch facilities. The Orbex family will be based on variants of a four-stage core, all-solid rocket launch vehicle. Orbex was designed to reuse Scout GSE.

The Orbex will be launched from LA-3 using the Scout erector/launcher. Processing of the launch vehicle stages will take place in the Orbex assembly building where the Orbex vehicle is integrated horizontally. This facility was previously used as the Scout Assembly Building. The payload is mated separately to the fourth stage, encapsulated then mated horizontally to the launch vehicle. The entire launch ready vehicle is transported to the launch pad where it is mated horizontally to the erector/launcher. Following final launch vehicle and payload preparations, the vehicle is rotated to the vertical position for launch. Launch operations are controlled from the blockhouse.

The Vehicle Assembly Building includes a class 10,000 clean room for spacecraft operations. Additional payload processing can be performed at existing NASA payload checkout and hazardous processing facilities on Wallops Island.

NASA Payload Processing Facilities

Payload Checkout and Assembly Area

The payload assembly and checkout facilities, provided for payload agency use, are temperature-controlled masonry buildings located midway between Launch Area No. 3 and the Dynamic Balance Facility. The Inert Bay and Office Building provides approximately 1,400 ft^2 of work space. The Hot Bay Building provides approximately 500 ft^2 of work space. The buildings are equipped with large doors, workbenches, office equipment, and chain fall hoists.

Dynamic Balance Facility

This facility, located on the north end of Wallops Island, provides a means for dynamically balancing the payload, its mounting hardware, adapters, and fourth-stage rocket motor. The facility consists of three buildings (a blockhouse and two test buildings), approximately 400 feet apart. The blockhouse contains a control room with all the necessary control, recording, and monitoring equipment to perform remote spin testing. Test Building No. 1 contains a Gisholt balancing machine capable of balancing up to 2,000 lb. Test Building No. 2 contains a Trebel balancing machine which is hard-bearing, permanently calibrated, and capable of balancing up to 6,000 lb. The fourth stage/payload assembly is mounted vertically on these machines, with 360° access provided by hydraulic platforms.

Point of Contact

NASA/Goddard Space Flight Center
Wallops Island, VA 23337
U.S.A.
Tel: 804-824-1579
Fax: 804-824-1683

United States, Western Range

T he Western Range (WR) launch facility is located at Vandenberg Air Force Base (VAFB) on the central coast of California at approximately 34°40' N., 120°37' W., about 50 miles west-northwest of Santa Barbara. It is situated on a prominent headland extending into the Pacific Ocean near Point Conception where the California coast changes from west- to south-facing. The VAFB coastline extends along 35 miles with both west- and south-facing beaches. The nearest major communities are Lompoc and Santa Maria, California, which are located 6 miles and 5 miles, respectively, from the base boundaries. The orientation of the coastline is ideal for high-inclination launch azimuths, including true polar and sun synchronous orbits. The range also includes instrumentation sites throughout the central California coast plus Hawaii, and can support suborbital launches, as well as mobile air and sea launches from the Pacific Ocean.

Vandenberg Air Force Base and the Western Range are owned by the U.S. Air Force. Occupied by the Chumash Indians for several thousand years, the VAFB area was essentially ranch land until just prior to World War II. At that time, the U.S. Army activated most of the current base as Camp Cooke for artillery training. In 1956, the U.S. Air Force acquired the northern portion of Camp Cooke for missile testing while the U.S. Navy operated a range from the south portion. By 1966, the Air Force had assumed all of the Navy assets as well as additional ranch land to reach its present size.

The base encompasses 98,265 acres and includes ICBM launch facilities along the west-facing coast of the northern portion of the base and space launch complexes near the south-facing coastline. Control centers and processing facilities are predominantly located in the central base area near the airfield. Instrumentation and communication sites are distributed throughout the base as well as at several offsite locations both up-range along the California coast and down-range in the Pacific. Polar space launches have the option of acquiring over-the-horizon telemetry using Advanced Range Instrumentation Aircraft (ARIA).

The topography of the more than 98,000 acres varies from coastal dunes to mountainous terrain. The maximum elevation on base is Mount Tranquillon at 2,160 feet. VAFB is in a seismically active earthquake zone, although severe earthquake activity has not been recorded in recent history. The large, varied, and extended beachfront and remote location give the VAFB Western Range many special features, which has resulted in its development into a multi-use launch facility. In addition to space launches, the

The Western Range launch facility is ideally situated for high-inclination launch azimuths, including true polar and sun synchronous orbits.

Western Range has conducted ballistic missile launches into the broad ocean area of the Pacific since 1958 and provides for a variety of aeronautical operations in the West Coast Offshore Operating Area. The first polar orbiting satellite in history was launched from VAFB in early 1959. Since then, over 540 high inclination and polar space launches have been conducted from the space launch complexes at Vandenberg. Launch azimuths for orbital launches are allowed between 147 and 201° or higher. Suborbital launches into the broad ocean area of the Pacific are allowed up to 281°.

There are presently six active, fixed launch complexes at VAFB from which space launches are conducted. These are Space Launch Complexes (SLC) 2W (Delta), 3W (Atlas E), 4E (Titan IV), 4W (Titan II), 5 (Scout), and Area 576E (Taurus). In addition, air-dropped space launches (Pegasus) are performed from the offshore aeronautical air space. Except for SLC-2W and Area 576E, all launch facilities are located on the southern portion of the base.

Both orbital and suborbital commercial space launches can be conducted from the Western Range as well as launches by the U.S. Government. Future capabilities are planned for several new launch vehicles. With the recent emergence of a state-chartered California Spaceport Authority, the new commercial space launch developments are expected to be multi-use facilities as opposed to the dedicated, uniquely operated complexes belonging to the U.S. Government. A new spaceport is being developed at SLC-6 and SLC-7 to accommodate these new launchers. Projected users of the spaceport concept include the Aquila, Conestoga, LLV, and Taurus. Commercial launches may also be available in 1996 using Atlas II launchers from SLC-3E. Future plans also include conversion of SLC-5 to serve commercial operations of the Orbital Express launch vehicle. Future suborbital launches are planned by AMROC to develop the components of the Aquila launch vehicle. Additional commercial payload processing facilities are being developed at both north and south VAFB locations to service a variety of users.

Launch Site Description

Delta Launch Operations

The Delta launch vehicle was first launched from the WR in 1966 by NASA. Since then, there have been 39 Delta launches at VAFB. The Delta launch facilities have recently completed modifications to accommodate the Delta II configurations. The facilities can accommodate two- and three-stage versions of the Delta II (6920/6925), as well as the more powerful strap-on solid motor configurations (7920/7925). First launch from the WR of the Delta II will occur in 1994. The prelaunch processing of the Delta launch vehicle takes place at facilities near SLC-2.

The SLC-2 consists of one launch pad (SLC-2W), a blockhouse, launch operations building, horizontal processing facility, and other facilities necessary to prepare, service, and launch the Delta vehicle. The Delta-II launch vehicle is fully integrated on the launch pad. The mobile service gantry and fixed umbilical tower were recently modified to accommodate the new, taller Delta II launch vehicle and larger strap-on solid rocket motors. The upper level enclosure, or white room area, provide protection for the spacecraft, checkout equipment, and personnel while on the launch pad. Launch operations are controlled from the blockhouse which is equipped with vehicle and ground systems monitoring and control equipment. A horizontal processing facility is located near SLC-2W for processing the Delta first and second stages. The facility is located south of the blockhouse and is used to perform prelaunch processing and preparations for erection and transport to the launch mount. A solid motor building is located near the SLC-2 area for processing of the strap-on solid rocket motors. The facility has been modified to handle both the Castor IV A and the larger graphite-epoxy motors. The facility is used for receiving, inspection, build-up, and preparation for erection and transportation to SLC-2W.

Payload processing is performed offpad at any of several processing sites. Integration with the Delta payload adapter/upper stage normally takes place at the NASA Spin Test Facility near SLC-2.

Pegasus Launch Operations

The Pegasus is a winged, three-stage, solid rocket booster that is air-launched from a carrier aircraft. The first Pegasus launch occurred at the WR in 1990. Since then, there have been three additional Pegasus launches, two of which took place from the WR. Orbital Sciences Corporation (OSC) builds and operates the Pegasus. A stretched version of the Pegasus (Pegasus XL) is in development with first launch planned for 1994. OSC has also procured a carrier aircraft to replace the aging B-52 used for development flights. The modified L1011 is based at Mojave, California. OSC activated a Vehicle Assembly Building (VAB) at the WR and will begin operations from the VAFB airfield in 1994.

All major vehicle subassemblies are delivered from the factory to the VAB. The Pegasus stages are integrated horizontally prior to arrival of the pay-load. The VAB has the capacity to integrate four Pegasus launch vehicles simultaneously. Integration is performed at a working height which allows convenient access to the vehicle for component installation, test, and inspection. A Pegasus payload is delivered to the integration site 7 to 21 days prior to launch. The payload must complete its own independent verification and checkout prior to beginning integrated processing with Pegasus at the VAB.

Following final close out, the Pegasus and payload are transported to the carrier aircraft for mating at the flight line. Preflight checks begin 4.5 hours before take-off. Flight and launch control functions are managed from the Launch Operations Control Room on the ground. Launch release functions are controlled from onboard the aircraft. At the WR, the Pegasus is typically launched over the Pacific Ocean about 50 miles west of the California coast. The range of achievable inclinations is 65 to 130°.

NASA Payload Processing Facilities

NASA Spacecraft Laboratory

The Spacecraft Laboratory is located in the Spacecraft Support Area on south VAFB. The Laboratory is a 53,859 ft^2 facility built in 1960 to support pro-cessing for small and medium polar orbit spacecraft. Upgrades performed in 1965 added a laboratory and microwave transmission capabilities. The two large laboratories with class 100,000 cleanliness ratings can accommodate spacecraft weighing up to 5 tons with diameters up to 10 feet. DS3, S- and X-band RF and video links are available for data transfer to the communications control building. Approximately 9,914 ft^2 of administrative space is available. No hazardous material provisions exist.

Scout Spin Test Facility

The Scout Spin Test facility is located on south VAFB. This hazardous processing facility is an 1,159 ft^2 block wall structure with steel siding whose primary function is spacecraft spin balance testing. Spacecraft up to 1,600 lb can be handled on the Gisholt vertical spin balance machine. Nearby buildings included in the compound area provide control and storage support. One high bay with a 1-ton bridge crane and one compressor room are in this facility. Telephone, 2-channel video, and RF data links are provided for communication. Approximately 200 ft^2 of administrative space is provided. Small rocket motor storage and pyrotechnic device handling are permitted at this facility.

Delta Spin Test Facility

The Delta Spin Test facility is located approximately 2 m (3.2 km) from SLC-2. A variety of spacecraft are processed in this 3,952 ft^2 facility built in 1966. An environmental equipment room was added in 1990. One class 100,000 clean-room-rated high bay is available that can handle spacecraft weighing up to 5 tons. The processing area is a class 100,000 clean room. The facility is rated for hazardous processing of ordnance, solid motors, and fuels. Hydrazine and oxidizer handling capabilities are available including a spill trench located in the high bay adjacent to the spin table. Data communications are provided via S- and X-band microwave, VHF, video, and operational intercoms. The spin balance table is capable of handling payloads weighing up to 6,000 lb. Both technical and facility groundings are available.

Orbex Launch Operations

CTA Launch Services plans to operate its Orbex launch vehicle from the WR using excess Scout processing and launch facilities. The Orbex family will be based on variants of a four-stage, core, all-solid rocket launch vehicle. Orbex was designed to reuse Scout GSE. First launch is planned in 1995.

The Orbex will be launched from SLC-5 using the Scout erector/launcher. Processing of the launch vehicle stages will take place in the Orbex Assembly Building where the Orbex vehicle is integrated horizontally. This facility was previously used as the Scout Ordnance Assembly Building. The payload is mated separately to the fourth stage, encapsulated, then mated horizontally to the launch vehicle. The entire launch-ready vehicle is transported to the launch pad where it is mated horizontally to the erector/launcher. Following final launch vehicle and payload preparations, the vehicle is rotated to the vertical position for launch. Launch operations are controlled from the SLC-5 blockhouse.

Lockheed Launch Vehicle Operations

Lockheed Missiles and Space Company, Inc., Missile Systems Division, along with United Technology Corporation Chemical Systems Division, Thiokol Corporation, and Rocket Research Company is developing a flexible family of launch vehicles that will be launched initially from the Western Range launch pad at Space Launch Complex 6 (SLC-6) and later from the Eastern Range. Their objective is to provide launch vehicles that meet individual spacelift requirements at substantial cost savings over today's alternatives without risking safety or reliability. The Lockheed Launch Vehicles (LLV) offer a modular approach: a series of incrementally higher performance vehicles and fairings. Configurations are tailored to specific mission requirements, allowing users to purchase only the capability needed. All configurations utilize the Thiokol Castor 120TM solid rocket motor as a first or second stage. The Orbus 21D®, a derivative of the United Technology (CSD) Orbus 21®, is used as a second or third stage. Rocket Research is providing the propulsion system for the orbit adjust module (OAM). The Thiokol Caster IVA is employed as a strap-on for the larger LLV3 configurations. Lockheed will flight test the first of the LLV family late in 1994.

Dynamic envelopes of the three payload fairings measure 78-in wide by 165-in high, 103-in wide by 269-in high, and 127-in wide by 298-in high. Payload weight capacity ranges from 2,343 to 8,980 lb.

California Commercial Space Launch Facility

The planned launch facility will be located at the site originally proposed as SLC-7 for Titan IV Centaur launches. The proposed complex encompasses approximately 135 acres near the southern extremity of VAFB on Cypress Ridge. Planned initial capability for the first launch pad is 1995. The complex will include three launch pads. The first and second launch pad designs are identical. The design will accommodate solid boosters such as the LLV, Taurus, and the hybrid Aquila. Although these boosters have different launch mount requirements, unique removable interface rings will adapt each configuration. Sounding rockets will use the third pad which is a flat, concrete launch pad. The exhaust duct uses a simple steel and concrete, modular, above-ground design to channel rocket exhaust away from the launch mount. The modular launch pad will be constructed of steel frames with reinforced concrete to make building blocks. Each of these blocks will be assembled at the launch site. Umbilical towers and access platforms for the various rocket designs will also be provided. The umbilical tower will provide critical services, such as electrical power and air conditioning, to the payload and booster. Access platforms will be similar to conventional scaffolding. The platforms will be covered to provide wind protection. A solid enclosure will also be provided at payload levels to allow access to payload areas on the launch pad. The complex also will offer an operations support building for personnel while launch vehicle and spacecraft processing occurs on the launch pad. Launch operations will be controlled from a launch control center located in the Integrated Processing Facility nearby at SLC-6.

Commercial Payload Processing Facilities

Astrotech Processing Facility

Astrotech Space Operations owns the facility which is being erected on land leased from VAFB. Astrotech operates the facility to the extent of providing space and capabilities for user organizations. Astrotech contracts with user organizations that process payloads according to their own requirements. Astrotech's primary objective is to build a commercial payload processing facility (PPF) that can process DOD, civil, and commercial payloads. Astrotech would provide a single interface for all payload launch site support requirements. The initial construction phase is planned for completion in 1994 and will provide facilities and capabilities required for the initial payload contract. Future construction will expand to the full processing capability and will provide the facilities for SLV/MLV payload processing.

Phase I will develop one cell. There will be two class 100,000 clean rooms. The clean room high bay will be 2,200 ft^2. The airlock clean room will be 2,400 ft^2. The facility will have a 700 ft^2 control room and a garment change room. The facility will meet explosion-proof standards and have a clean room storage area. A 10-ton bridge crane will be installed with a 37-foot hook height. Administrative support will be temporarily housed nearby until the completion of Phase II construction. Phase II will complete two cells as facilities with bipropellant fueling capabilities. These facilities will include a dynamic spin balancing machine. The second cell will duplicate the first but with a taller airlock.

California Commercial Space Integrated Processing Facility

The California Commercial Integrated Processing Facility (IPF) is planned for availability in 1994. The building is part of Space Launch Complex 6 (SLC-6) which was built in 1987 to provide a processing capability for the Space Transportation System payloads. It is a 103,830 ft^2 reinforced concrete structure containing three checkout cells, each with seven service levels and access from an erection room that provides lifting service from a 75-ton bridge crane. Cell dimensions are 35 feet by 44 feet by 69 feet, and each has an access door area of 23 feet by 72 feet. Payload access to the facility is through a horizontal airlock also containing two 25-ton cranes. The entire processing area is clean room rated at class 100,000. The checkout cells each provide a 5-ton bridge crane for local lifting services. Piping systems are installed to handle hydrazine propellants and spill containment. Both facility and isolated single-point technical grounding is provided throughout.

Personnel and equipment access to the service cells is available through both clean and standard elevators. The administrative section provides offices, equipment storage, clean room garment changing, storage areas, and personnel ingress/egress. Both electronic and physical security capabilities exist for personnel access. A transfer tower exists in this facility containing a 75-ton hoist. This crane can lift fairing components or a payload to the upper transfer level for additional operating space. The IPF will be used as a combined payload and booster processing facility, a fairing processing and storage facility, a payload encapsulation facility, and a launch control center.

Point of Contact

30th Space Wing/XP
  747 Nebraska Ave., Suite 34
  Vandenberg Air Force Base, CA 93437-6294
  U.S.A.
  Tel: 805-734-8232, Ext. 67363
  Fax: 805-734-8232, Ext. 68608

Appendix 1

Commercial Launch Vehicles

Commercial Launch Vehicles—Geosynchronous Transfer Orbit (GTO)

Launch Vehicle	Payload Capacity, kg	Mission Option(s)	Launch Site Location(s)
Current Operational			
Proton	5,700 (GTO) 2,600 (GEO)	Dedicated payload to GTO or GEO	Kazakhstan Baikonur Cosmodrome
Ariane 44L	4,460	Dedicated or shared (dual) payloads	French Guiana Guiana Space Center
Commercial Titan III	4,320+	Dedicated or shared (dual) payloads	United States Eastern Range
Ariane 44LP	4,030	Dedicated or shared (dual) payloads	French Guiana Guiana Space Center
Atlas IIAS	3,500	Single payload	United States Eastern Range
Long March 2E	3,460	Dedicated or shared (dual) payloads	China Xichang Launch Center
Ariane 42L	3,350	Dedicated or shared (dual) payloads	French Guiana Guiana Space Center
Ariane 44P	3,290	Dedicated or shared (dual) payloads	French Guiana Guiana Space Center
Atlas II	2,810	Single payload	United States Eastern Range
Ariane 42P	2,765	Dedicated or shared (dual) payloads	French Guiana Guiana Space Center
Atlas II	2,680	Single payload	United States Eastern Range
Long March 3A	2,500	Single payload	China Xichang Space Center
Ariane 40	2,020	Dedicated or shared (dual) payloads	French Guiana Guiana Space Center
Delta II/7925	1,820	Single payload	United States Eastern Range
Delta II/6925	1,450	Single payload	United States Eastern Range
Long March 3	1,500	Single payload	China Xichang Space Center
Pegasus I	125	Air-launched Single payload	United States (To be determined)

Commercial Launch Vehicles—Geosynchronous Transfer Orbit (GTO) (continued)

Launch Vehicle	Payload Capacity, kg	Mission Option(s)	Launch Site Location(s)
In Development—Near Term (1994–1998)			
Ariane 5	6,800 (dedicated) 5,900 (dual) 5,500 (triple)	Dedicated or shared (dual, triple) payloads	French Guiana Guiana Space Center
H-II	4,000	Dedicated or shared (dual) payloads	Japan Tanegashima Space Center
Taurus	430	Single payload	United States (To be determined)
Proposed Developments (1995–2000+)			
Energiya-M	13,000	Single or multiple payloads	Kazakhstan Baikonur Cosmodrome
Proton	(GTO) 4,000 (GEO)	Single payload to GTO or GEO	New Guinea Papua Space Port
Proton-KM	7,800 (GTO) 3,500 (GEO) 6,000 (GTO)	Single payload to GTO or GEO Single payload to GTO	Kazakhstan Baikonur Cosmodrome Russia Plesetsk Cosmodrome
Long March 3B	4,800	Single or multiple payloads	China Xichang Space Center
Zenit 3	3,800	Single payload	Kazakhstan Baikonur Cosmodrome
	5,000	Single payload	Australia
Taurus XL, XL/S	520, 690	Single payload	United States (To be determined)

Commercial Launch Vehicles—Medium-Lift Low Earth Orbit (MLEO) (Payload Capacity Less Than 10,000 kg)

Launch Vehicle	Payload Capacity, kg	Mission Option(s)	Launch Site Location(s)
Current Operational			
Long March 2E	9,460	Inclinations: 28.5° 60.0°	China Xichang Space Center Jiuquan Space Center
Atlas IIAS	3,500	Single payload	United States Eastern Range
Long March 3A	2,300	Single payload	China Xichang Space Center
Ariane 42L, 44LP, 44L	7,000+	Inclinations above 5.2° SSO	French Guiana Guiana Space Center
Soyuz	7,500	(Man-rated) Inclinations above 51°	Kazakhstan Baikonur Cosmodrome
	7,200	Inclinations above 63°	Russia Plesetsk Cosmodrome
Ariane 44P	6,900	Inclinations above 5.2° SSO	French Guiana Guiana Space Center
Atlas IIA	6,760	Inclinations of 39 to 57°	United States Eastern Range
Atlas II	6,580	Inclinations of 39 to 57°	United States Eastern Range
Vostok	6,300	SSO mission only	Kazakhstan Baikonur Cosmodrome
Ariane 42P	6,000	Inclinations above 5.2° SSO	French Guiana Guiana Space Center
Delta II/7920	5,040	Inclinations of 39 to 57°	United States Eastern Range
Long March 3	5,000	Inclinations of 29 to 33°	China Xichang Space Center
Ariane 40	4,900	Inclinations above 5.2°; SSO	French Guiana Guiana Space Center
Long March 4	4,000	SSO mission only	China Taiyuan Space Center
Delta II/6920	3,980	SSO mission only	United States Western Range
Tsyclon	3,500	Inclinations of 74°, 83°	Russia Plesetsk Cosmodrome
Long March 2D	3,100	Payload recovery capsule. Inclinations above 42°	China Jiuquan Space Center
Long March 2C	2,500	Inclinations above 42°	China Jiuquan Space Center

73

Commercial Launch Vehicles—Medium-Lift Low Earth Orbit (MLEO) (Payload Capacity Less Than 10,000 kg) (continued)

Launch Vehicle	Payload Capacity, kg	Mission Option(s)	Launch Site Location(s)
In Development—Near-Term (1994–1998)			
Atlas IIAS	8,400	Inclinations of 39 to 57°	United States Eastern Range
Rus (Soyuz)	8,300	Inclinations above 51°	Kazakhstan Baikonur Cosmodrome
	7,800	Inclinations above 63°	Russia Plesetsk Cosmodrome
Long March 3A	7,500	Inclinations of 29 to 33°	China Xichang Space Center
Atlas IIAS	6,800	SSO mission only	United States
Atlas IIA	5,500	SSO mission only	Western Range
VITA/SS-18K	4,000	Payload recovery capsule; inclinations above 51°	Kazakhstan Baikonur Cosmodrome
Polar SLV	3,000/LEO 1,000/SSO	46° inclination, SSO	India Srihirakota Rocket Launching Station
Proposed Developments (1995–2000+)			
Lockheed Launch Vehicle 3 (LLV3)	4,080	(Unknown)	United States Eastern Range Western Range

Commercial Launch Vehicles—Medium-Lift Low Earth Orbit (MLEO) (Payload Capacity More Than 10,000 kg)

Launch Vehicle	Payload Capacity, kg	Mission Option(s)	Launch Site Location(s)
Current Operational			
Energiya	100,000	Inclinations above 51° (51–83°, 97°)	Kazakhstan Baikonur Cosmodrome
Proton	22,000	Inclinations above 51°	Kazakhstan Baikonur Cosmodrome
Commercial Titan III	14,500	Inclinations of 39 to 57°	United States Eastern Range
Zenit 2	13,500	Inclinations above 51°; SSO	Kazakhstan Baikonur Cosmodrome
In Development—Near-Term (1994–1998)			
Ariane 5	18,000 (LEO) 10,000 (SSO)	Inclinations above 5.2° SSO	French Guiana Guiana Space Center
H-II	10,000	Inclinations above 31°	Japan Tanegashima Space Center
Proposed Developments (1995–2000+)			
Energiya-M	35,000	Inclinations above 51°	Kazakhstan Baikonur Cosmodrome
Proton	24,000 (LEO) 17,000 (SSO)	Inclinations above 2° SSO	New Guinea Papua Space Port
Proton-KM	23,700	Inclinations above 51°	Kazakhstan Baikonur Cosmodrome
	22,300	Inclinations above 63°	Russia Plesetsk Cosmodrome
Zenit 2	15,000	Inclinations above 12.5°	Australia Cape York Space Port
Long March 3B	13,500	Inclinations above 29 to 33°	China Xichang Space Center
Zenit 2	12,700	Inclinations above 63°	Russia Plesetsk Cosmodrome

Commercial Launch Vehicles—Light-Lift Low Earth Orbit (LLEO)
(Payload Capacity Less Than 2,500 kg)

Launch Vehicle	Payload Capacity, kg	Mission Option(s)	Launch Site Location(s)
Current Operational			
Cosmos	1,500	Inclinations: 66°, 74°, 83° above 51°	Russia Plesetsk Cosmodrome Kapustin Yar Cosmodrome
Long March 1D	750	Inclinations above 42°	China Jiuquan Space Center
Augmented SLV	150	——	India Srihirakota Rocket Launching Station
In Development—Near-Term (1994–1998)			
Taurus	1,360	——	United States Eastern Range Western Range Wallops Flight Facility
Lockheed Launch Vehicle 1 (LLV1)	1,040	——	United States Eastern Range Western Range
SS-N-20 (derivative)	1,000	——	Russia Unidentified
Comet – Conestoga 1620	820	Recovery capsule	United States Wallops Flight Facility
START-1	759	——	Russia Plesetsk Cosmodrome
Lockheed Launch Vehicle 2 (LLV2)	2,400	——	United States Eastern Range Western Range
M-5	1,800	——	Japan Kagoshima Space Center
Southern Launch Vehicle	<1,500	——	Australia Woomera Rocket Range
Conestoga II–IV	226–1,400	——	United States Wallops Flight Facility
Vega	600–1,000	——	Italy San Marco Rangeln

Commercial Launch Vehicles—Light-Lift Low Earth Orbit (LLEO)
(Payload Capacity Less Than 2,500 kg)
(continued)

Launch Vehicle	Payload Capacity, kg	Mission Options(s)	Launch Site Location(s)
Development—Near-Term (1994–1998)			
J-1	900	——	Japan Tanegashima Space Center
SLV (Sweden)	250	High inclination missions	Sweden Esrange
Capricornio	220	——	Spain Canary Islands
VLS (Brazilian Satellite Launch Vehicle)	200	Variety of inclinations above 2°	Brazil Alcantara Launch Center
Orbital Express	200	——	United States Wallops Flight Facility Poker Flat Research Range
SLV (Norway)	150	High inclination missions	Norway Andoya Rocket Range

Commercial Launch Vehicles—Transportable/Mobile Low Earth Orbit (TMEO) (Payload Capacity Less Than 2,500 kg)

Launch Vehicle	Payload Capacity, kg	Mission Options(s)	Launch Site Location(s)
Current Operational			
Pegasus I (air-launched)	250–450	——	United States Current launch zones: Western Range Eastern Range Wallops Flight Facility
In Development—Near-Term (1994–1998)			
Surf (sea-launched)	2,400	Floating water-launch technique	Russia (To be determined)
Taurus (transportable, land-launched)	1,360	Transportable to prepared launch positions	United States (To be determined)
Burlak (air-launched)	1,100	——	Russia (To be determined)
START-1 (mobile transporter)	750	Transportable to customer launch location	Russia (To be determined)
Pegasus II (air-launched)	750	——	United States (Launch ranges currently supporting Pegasus I)
SHTIL 3 (sea-launched)	670	Floating water-launch technique	Russia (To be determined)
Pegasus IB (air-launched)	600	——	United States (Launch ranges currently supporting Pegasus I)
SHTIL/SS-N-23 (submarine launched ?)	550–750	——	Ukraine (Broad ocean areas)
SHTIL 1 (sea-launched)	430	Floating water-launch technique	Russia (To be determined)
SHTIL 2 (sea-launched)	265	Floating water-launch technique	Russia (To be determined)
Volna/SS-N-18 (submarine launched ?)	125	——	Ukraine (Broad ocean areas)
Vysota/SS-N-8 (submarine launched ?)	115	——	Ukraine (Broad ocean areas)
Proposed Developments (1995–2000+)			
Pegasus III, IV (air launched)	1,225–3,100	——	United States (Launch ranges currently supporting Pegasus I)
SHTIL 3A (air-drop launch)	Unknown	——	Ukraine (To be determined)
Space Clipper (6 versions) (air-drop launched)	500–2,200	——	Ukraine (To be determined)

Commercial Launch Vehicles/Services—Suborbital, Sounding Rockets (SSR)

Launch Vehicle	Payload Capacity, kg	Mission Option(s)	Launch Site Location(s)
Current Operational			
Sounding Rocket Services—Sweden	——	Universal launchers; altitudes, to 1,000 km; payload recovery (land impact area)	Sweden Esrange—Sounding Rocket and Balloon Facility
Sounding Rocket Services—Norway	——	Universal launchers; altitudes to 800 km; 4-stage rockets (up to 20 ton class); 1,900 km (max) impact range	Norway Andoya Rocket Range
TR-1A	750	Altitude to 290 km; microgravity	Japan Tanegashima Space Center Takesaki Range
SONDA III	60	Altitude to 600 km	Brazil Alcantara Launch Site
Sounding Rocket Services Pakistan	——	(Open to negotiation)	Pakistan SUPARCO Flight Test Range
Centaure; Dragon	60	Altitude to 440 km	
In Development—Near-Term (1994–1998)			
SS-N-18 (sea-launched)	(Unknown)	Suborbital; microgravity time = 50 min	Russia (To be determined)
SS-N-8 (sea-launched)	(Unknown)	Suborbital; microgravity time = >20 min	Russia (To be determined)
SS-N-6	(Unknown)	Suborbital; microgravity time = >17 min	Russia (To be determined)
MAXUS	(350)	Altitude to 1,000 km; microgravity time = 15 min	Sweden Esrange—Sounding Rocket and Balloon Facility
SONDA IV	(500)	Altitude to 700 km	Brazil Alcantara Launch Site
Proposed Developments (1995–2000+)			
Sounding rocket (unidentified)	(Unknown)	High altitude research	China Hainan Island Launch Center

Appendix 2

Worldwide Launch Sites and Launch Service Offerings

Appendix 2 provides a comprehensive listing of historical, operational, and proposed launch sites worldwide, along with their corresponding Launch Vehicle and mission offerings.

The abbreviations used in Appendix 2 are as follows:

GEO	Geosynchronous Orbit (Full Service)
GTO	Geosynchronous Transfer Orbit
HEO	Highly-Elliptical Orbit/12-hour
HLEO	Heavy-Lift Low Earth Orbit
LLEO	Light-Lift Low Earth Orbit
MLEO	Medium-Lift Low Earth Orbit
RECOV	Payload Recovery Capsule
SSO	Sun Synchronous/Polar Orbit
SSR	Suborbital Sounding Rocket
TMLEO	Transportable/Mobile Low Earth Orbit

* Asterisk items are described in the main body of this book.

Worldwide Launch Sites and Launch Service Offerings

Country of Origin/ Site Name	Launch Vehicle(s)	Mission/ Offering(s)	Commercial Status
Argentina			
Mar Chiquita Launch Site	Unknown	SSR/none to date	Unknown
La Rioja Launch Site	Unknown	SSR/none to date	Unknown
Australia			
Cape York Space Port	Zenit 2/3	GTO, HLEO	Proposed
Woomera Rocket Range*	Optional	SSR, LLEO	Operational
Brazil			
Alcantara Launch Center*	VLS	LLEO	In development
	SONDA II	SSR	Operational
	SONDA III	SSR	Operational
Barreira do Inferno Launch Center*	SONDA II	SSR	Operational
	SONDA III	SSR	Operational
	SONDA IV	SSR	In development
Canada			
Spaceport Canada*	Unknown	SSR, LLEO	In development

Worldwide Launch Sites and Launch Service Offerings (Continued)

Country of Origin/ Site Name	Launch Vehicle(s)	Mission/ Offering(s)	Commercial Status
China			
Xichang Satellite Launch Center*			
Launch Pad No. 1	Long March 3	GTO	Operational
Launch Pad No. 2	Long March 2E	GTO	Operational
	Long March 3A	GTO	Operational
	Long March 3B	GTO	In development
Jiuquan Satellite Launch Center*	Long March 1D	LLEO	Operational
	Long March 2C	MLEO	Operational
	Long March 2D	MLEO, RECOV	Operational
Taiyuan Satellite Launch Center*	Long March 4	MLEO/SSO	Operational
Hainan Island Launch Center	Unknown	SSR/none to date	Proposed
France (Kourou, French Guiana)			
Guiana Space Center*			
Ensemble de Lancement Ariane 2 (ELA 2)	Ariane 4 (6 versions)	GTO, MLEO, SSO	Operational
Ensemble de Lancement Ariane 3 (ELA 3)	Ariane 5	GTO, HLEO, SSO	In development
Ensemble de Lancement Ariane 1 (ELA 1)	Deactivated	Not applicable	Not available
India			
Srihirakota Range*	Augmented SLV	LLEO	Proposed
	Polar SLV	(SSO, MLEO)	In development
	Geosynch SLV	GTO/none to date	In development
	Unidentified	SSR/none to date	Unknown
Balasore Rocket Launching Station	Unidentified	SSR/none to date	Unknown
Thumba Rocket Launching Station	Unidentified	SSR/none to date	Unknown
Indonesia			
Pameungpeuk Range	Unknown	None to date	Unknown
Israel			
Palmachim—Shavit Launch Site	Shavit	LLEO/none to date	Not available
Italy			
San Marco Equatorial Range*	Vega	LLEO	In development
Japan			
Tanegashima Space Center*			
Yoshinobu Launch Site	H-II	GTO, HLEO	In development
Osaki Launch Site	J-1	LLEO	In development
Takesaki Launch Site	TR-1A	SSR	Operational
Kagoshima Space Center*	M-5	LLEO/none to date	Proposed
Kazakhstan			
Baikonur (Tyruatam) Cosmodrome*			
Energiya Launch Sites	Energiya (Russia)	HLEO	Proposed
	Energiya-M (Russia)	GTO, HLEO	In development
Proton Launch Sites	Proton (Russia)	GEO, GTO, HLEO	Operational
	Proton-KM (Russia)	GEO, GTO,	In develoment
Zenit Launch Site	Zenit 2 (Ukraine)	HLEO	Operational
	Zenit 3 (Ukr./Russ.)	HLEO	In development
Soyuz Launch Sites	Soyuz (Russia)	GTO	Operational
	Vostok (Russia)	MLEO	Operational
	Molniya (Russia)	MLEO/SSO	Operational
	Rus (Russia)	HEO	Unknown
Unidentified Site	VITA/SS-18K (Ukr.)	MLEO/none	Proposed
Unidentified Site	Rokot (Russia)	MLEO, RECOV LLEO	Proposed

Worldwide Launch Sites and Launch Service Offerings
(continued)

Country of Origin/ Site Name	Launch Vehicle(s)	Mission/ Offering(s)	Commercial Status
Norway			
Andoya Rocket Range*	SLV	LLEO	Proposed
	Sounding rockets	SSR services	Operational
Pakistan			
SUPARCO Flight Test Range*	Centaure; Dragon	SSR services	Open to negotiation
Papua New Guinea			
Space Port	Proton	GEO, GTO, HLEO, SSO	Proposed
Russia			
Baikonur (see Kazakhstan)*			
Plesetsk (Northern) Cosmodrome*	Soyuz	MLEO, RECOV	Operational
Soyuz Launch Sites	Molniya	HEO	Operational
	Rus	None to date	In development
	Tsyclon (Ukraine)	MLEO	Operational
Tsyclon Launch Site	Cosmos (Ukraine)	LLEO	Operational
Cosmos Launch Site	START-1	LLEO, TMLEO?	In development
START-1 Launch Site	Zenit (Ukraine)	None to date	Not available
Zenit Launch Site	Proton	None to date	Proposed
Proton Launch			
Kapustin Yar Cosmodrome	Cosmos (Ukraine)	Uknown	Unknown
Cosmos Launch Site	SS-N-20	LLEO	Proposed
(Unidentified site)			
Transportable Launch System			
Air-launched	Burlak	TMLEO	In development
	SHTIL 3A	TMLEO	Proposed
Sea-launched	Surf	TMLEO	In development
	SHTIL 3	TMLEO	Proposed
	SHTIL 2	TMLEO	Proposed
	SHTIL 1	TMLEO	Proposed
	SS-N-18	SSR	Proposed
	SS-N-8	SSR	Proposed
	SS-N-6	SSR	Proposed
Land-mobile	START-1	TMLEO	Proposed
Spain			
Canary Islands (proposed)	Capricornio	LLEO	In development
Sweden			
Esrange – Sounding Rockets and	SLV	LLEO	Proposed
Balloon Facility	MAXUS	SSR	In development
	Sounding Rockets	SSR services	Operational
Taiwan			
PingTung	Unknown	Unknown	Not available

Worldwide Launch Sites and Launch Service Offerings
(continued)

Country of Origin/ Site Name	Launch Vehicle(s)	Mission/ Offering(s)	Commercial Status
Ukraine			
Baikonur (see Kazakhstan)*	——	——	——
Plesetsk (see Russia)	——	——	——
Kapustin Yar (see Russia)	——	——	——
Transportable Launch Systems			
Air-launched	Space Clipper/ 6 versions	TMLEO	Proposed
Sea-launched	SHTIL	LLEO	Proposed
	Volna	LLEO	Proposed
	Vysota	LLEO	Proposed
United States			
Cape San Blas Launch Operations*			
Microstar Launch Site	Microstar	SSR	Operational
Eastern Range (ER)*			
Atlas Launch Sites			
Delta II Launch Sites	Atlas II (3 versions)	GTO, MLEO	Operational
Commercial Titan Launch Site	Delta II (2 versions)	GTO, MLEO	Operational
Taurus Launch Site	Commercial Titan III	GTO, HLEO	Operational
	Taurus	LLEO	Proposed
Lockheed Launch Vehicle Launch Site	Taurus XL, XL/S	LLEO, GTO?	Proposed
	LLV1, 2	LLEO	Proposed
Western Range (WR)*	LLV3	MLEO	Proposed
Atlas Launch Site			
Delta II Launch Site	Atlas II (3 versions)	MLEO, SSO	Not available
Taurus Launch Site	Delta II/7920	MLEO, SSO	Operational
	Taurus	LLEO	In development
Lockheed Launch Vehicle Launch Site	Taurus XL, XL/S	LLEO	Proposed
	LLV1, 2	LLEO	Proposed
Wallops Flight Facility*	LLV3	MLEO	Proposed
Conestoga Launch Site	LLV3	MLEO	Proposed
Taurus Launch Site	Conestoga (3 vers.)	LLEO	In development
	Taurus	LLEO	Proposed
Orbital Express Launch Site	Taurus XL, XL/S	LLEO	Proposed
Poker Flat Research Range*	Orbital Express	LLEO	Proposed
Orbital Express Launch Site			
Sounding Rocket Launch Site(s)	Orbital Express	LLEO	Proposed
	Unknown	SSR	Proposed
Transportable Launch Systems			
Air-launched	Pegasus I	TMLEO, GTO	Operational
Dryden Flight Research Facility #	Pegasus IB	TMLEO, GTO	Proposed
# (WR, ER, WFF provided range support)	Pegasus I, II, III, IV	TMLEO, GTO	Proposed
Land-launched	Taurus	TMLEO, GTO	Proposed
	Taurus XL, XL/S	TMLEO, GTO	Proposed

Appendix 3

Launch Site Mailing Addresses (By Country)

ARGENTINA

Instituto de Investigaciones
  Technológicas de la Fuerza Aérea
  Avenida Fuerza Aérea Argentina Km 5. 1/2,
  5103 Cordoba, Argentina

Comision Nacional de
  Actividades Espaciales (CONAE)
  Edificio Eolo, Avenida Dorrego 4010,
  1425 Buenos Aires, Argentina

AUSTRALIA

Australia Space Office
  P.O. Box 269 Civic Square
  Canberra, ACT. 2608
  Australia
  Tel: 61-6276-1490
  Fax: 61-6276-1567

Economic Development Authority
  Government of South Australia
  GPO Box 1264
  Adelaide 5001
  South Australia
  Attn: Nigel F. Barkham
  Tel: 61-8210-8622

BRAZIL

Centro de Lançamento de Alcantara
  CEP 65.250-000–Alcantara–MA, Brasil
  Fax: 55-98-2211069

Centro de Lançamento de Barreira do Inferno
  RN 063, KM 11
  Caixa Postal 640
  59.150-000 Natal–RN, Brasil
  Fax: 55-84-2114266

Departamento de Pesquisas y
  Desenvolvimento
  Esplanada dos Ministérios—
  Bloco M
  Edifício Anexo do MAer, 3° Andar
  70.045-900–Brasilia DF
  Brasil
  Fax: 55-61-2246123

Instituto de Aeronáutica e Espaço
  CP 6001, 12225 São José dos Campos,
  São Paulo, Brasil
  Fax: 55-123-412522

CANADA

Canadian Space Agency
  Place Air Canada
  500 René-Lévesque Boulevard West
  Montreal, Quebec H2Z 1Z7, Canada

Manitoba Aerospace Technology Program
  500-15 Carlton Street
  Winnipeg, Manitoba R3C 3H8
  Tel: 1-204-945-2030
  Fax: 204-957-1793/945-1354

CHINA

China Great Wall Industry Corporation
  22 Fucheng Road
  Beijing 100036, China

China Great Wall Industry Corp. (GWIC)
  21 HuangSi DaJie
  Xicheng Qu
  Beijing, 100011
  P.R. China
  Tel: 861-837-2708, 861-837-1682
  Fax: 861-837-3155, 861-837-2693

GW Aerospace, Inc.
  21515 Hawthorne Blvd. #1065
  Torrance, CA 90503
  U.S.A.

EUROPEAN SPACE AGENCY/FRANCE
FRENCH GUIANA

Arianespace, Inc.
  700 13th Street, NW, Suite 200
  Washington, D.C. 20005
  U.S.A.
  Attn: Michelle Lyle

BP-726, F-97387 Kourou Cedex
  French Guiana
  Tel: 594-33-51-11
  Fax: 594-33-47-66

INDIA

Indian Space Research Organisation
  (ISRO)
  Anatariksha Bhavan,
  New BEL Road
  Bangalore 560 094, India

Srihirakota Range
  Andhra Pradesh 524124
  India
  Tel: 91-2001-041-394
  Fax: 91-2001-041-568594

INDONESIA

National Institute of Aeronautics and Space
  JL Pemuda, Persil No. 1
  P.O. Box 20/JAT
  Jakarta 13220, Indonesia

ISRAEL

Israel Space Agency
  P.O. Box 17185
  26a Chaim Levanon Street
  Ramat-Aviv, 61171 Tel Aviv, Israel

ITALY

Agenzia Spaziale Italiana (ASI)
  Via di Villa Patrizi 13,
  00161 Rome, Italy
  Tel: 39-6-440-4205
  Fax: 39-6-440-4212

Agenzia Spaziale Italiana (ASI)
  250 E Street, SW, Suite 30
  Washington, D.C. 20024
  U.S.A.
  Attn: Mr. Enzo Letico

JAPAN

Institute of Space and Astronautical
  Science (ISAS)
  3-1-1 Yoshinodai
  Sagamihara-shi
  Kanagawa-ken 229, Japan

Kagoshima Space Center
  Uchinoura, Kimotsuki-gun
  Kagoshima 893-14
  Japan
  Tel: 81-99-467-2211
  Fax: 81-99-467-3811

National Space Development Agency of
  Japan
  633 West Fifth Street, Suite 5870
  Los Angeles, CA 90071, U.S.A.
  Attn: Mr. Shizuo Hoshiba
  Tel: 213-688-7758
  Fax: 213-688-0852

Rocket System Corporation
  Sumitomo Shibadaimon Bldg., 12F
  2-5-5 Shibadaimon, Minato-Ku
  Tokyo 105, Japan

Tanegashima Space Center
  Minamitane-machi, Kumage-gun
  Kagoshima 891-37
  Japan
  Tel: 81-99-726-2111
  Fax: 81-99-726-0199

KAZAKHSTAN

The Space Commerce Corporation
  5718 Westheimer, Suite 1515
  Houston, TX 77057
  U.S.A.
  Tel: 713-977-1543

NORWAY

Norsk Romsenter (Norwegian Space
  Center)
  Hoffsveien 65A
  P.O. Box 85 Smestad
  N-0309 Oslo 9, Norway
  Tel: 47-2-52-38-00
  Fax: 47-2-23-97

PAKISTAN

Pakistan Space and Upper Atmosphere
  Research Commission (SUPARCO)
  Sector 28, Gulzar-e-Hijri
  Off University Road, P.O. Box 8402
  Karachi 75270, Pakistan
  Attn: M. Nasim Shah
  Tel: 474261-4
  Fax: 92-21-496-0553

RUSSIA

The Space Commerce Corporation
  5718 Westheimer, Suite 1515
  Houston, TX 77057
  U.S.A.
  Tel: 713-977-1543

Glavkosmos
  9, Krasnoproletarskaya St
  Moscow, 103030
  Russia
  Attn: Alexander Dunaev
  Tel: 972-44-97
  Fax: 288-95-83

85

SOUTH AFRICA

Houwteq
  Private Bag X8
  GRABOUW
  7160 South Africa
  Attn: Mr. R. Balt
  Tel: 27-24-50-5911
  Fax: 27-24-59-31-67

SPAIN

Instituto Nacional de Técnica
  Aeroespacial (INTA)
  Carretera de Torrejon-Ajaluir, km 4
  Torrejon de Ardoz
  28850 Madrid, Spain

SWEDEN

Swedish Space Corporation
  P.O. Box 4207, Albygatan 107
  S-171 04 Solna, Sweden

UNITED STATES

Cape San Blas Launch Operations
  Spaceport Florida Authority
  Cocoa Beach, FL, U.S.A.
  Attn: Edward Ellegood
  Tel: 407-868-6983

Eastern Range
  45th Space Wing/XP
  Patrick Air Force Base, FL, U.S.A.
  Tel: 407-494-5938
  Fax: 407-494-7302

Poker Flat Research Range
  Geophysical Institute
  University of Alaska, Fairbanks
  Fairbanks, AK 99775-0800, U.S.A.
  Tel: 907-474-7413
  Fax: 907-474-7015

Wallops Flight Facility
  NASA/Goddard Space Flight Center
  Wallops Island, VA 23337, U.S.A.
  Tel: 804-824-1579
  Fax: 804-824-1683

Western Range
  30th Space Wing/XP
  747 Nebraska Ave., Suite 34
  Vandenberg Air Force Base
  CA 93437-6294, U.S.A.
  Tel: 805-734-8232, Ext. 67363
  Fax: 805-734-8232, Ext. 68608

UKRAINE

Yuzhkosmos
  3 Krivorozhskaya Street
  320008 Dnepropetrovsk, Ukraine

Bibliography

Andrew Wilson, ed., *Jane's Space Directory 1993–1994,* Ninth Edition, 1993, Biddles Ltd, Guildford, and King's Lynn.

The Illustrated Encyclopedia of Space Technology, Second Edition, 1989, Salamander Books Limited.

Steven J. Isakowitz, *International Reference Guide to Space Launch Systems,* 1991 Edition, American Institute of Aeronautics and Astronautics.

Frank Sietzen Jr., *World Guide to Commercial Launch Vehicles,* 1991, Pasha Publications.

World Space Systems Briefing, 1993, Teal Group Corporation, Briefing Book Series.

Index